First Year Calculus An Inquiry Based Learning Approach

Clement E. Falbo

FolioAvenue Publishing Service
2031 Union Street, Suite 6, San Francisco CA 94123
415-869-8834 (866-365-4628)
www.folioavenue.com

First Year Calculus, an Inquiry-Based Learning Approach
Copyright © 2019 Clement E. Falbo

Printed in the United States of America

ISBN: Paperback :   978-1-949473-68-1
       Hardcover :   978-1-949473-80-3

Dedication

I thank my wife, Jean, for proofreading every word of this book, more than once and for suggesting clarifying changes. If any errors slipped by they were is places that I changed after she had read and corrected the manuscript for the last time.

Contents

CONTENTS vii

Preface

We owned the calculus!

This book is called First Year Calculus, An Inquiry Based Learning (IBL) Approach because it progresses through the topics of calculus by questions and answers, a process akin to the Socratic Method initiated 2400 years ago. It worked then and it works really well today. But, what is IBL and how does it relate to this calculus book I have written for you?

Throughout the decades from the 1930s to the 1970s, Dr. R. L. Moore and his colleagues in the Pure Mathematics Department at the University of Texas, Austin were using a very unusual way to teach mathematics. No lectures.

These professors taught their classes by having students take their turns at becoming the teachers and presenting their solutions to difficult homework problems in class in front of their fellow students. This approach was known among various colleges and universities in the United States, and even internationally as the Moore Method, or the Texas Method.

We students at Texas experienced the excitement of being able to solve hard problems, and being recognized for our accomplishment by being the person who got to explain the solution to our classmates. More importantly, we realized that when we explained a concept, we really learned the subject much better than just hearing someone lecture about it. We owned the calculus, and topology, and differential equations, and other mathematical subjects!

Over the years, students who were taught like this emerged as mathematicians publishing their research and teaching mathematics by their own student--centered techniques, often called Modified Moore Methods. Unlike Dr. Moore himself, these practitioners do use textbooks, often to furnish their students with definitions and examples, but they still retain the process of having students teach students in the presence of the teacher.

The amalgam that arose from these various approaches has become, what is now known as IBL, a major way that mathematics is being taught in American colleges today.

As with the Socratic method, a big part of the success of IBL comes from selecting the right kind of questions to ask and the right kind of problems to pose. I characterize them as "problems that teach" because their solutions provide some insight and understanding as well as some advancement of the subject.

Students who want to learn calculus on their own may do so from this book. The material here is, essentially, a transcription of the notes I took in the 1955-1956 semester while I was a student in Dr. Moore s class. I have included all of his problems (problems that teach,) in the text and I have deliberately included all of the solutions we got in our class. Readers can compare their answers to the ones we got. The definitions, discussions, examples are about the same as we encountered in class including some cases in which we ran into blind alleys, false starts and restarts on key problems.

In a later section of this preface, I give you an outline on "how to use this book." Any student who is learning calculus and wants to experience the satisfaction of discovering and "owning" calculus can do so by following this outline.

This book is designed for classroom use by any teacher who is interested in teaching calculus with a strong student-centered approach. A teacher using this book must be willing to be an "interested bystander" while the students are the ones who present solutions at the board. Instead of lecturing, the teacher might ask for clarification or call on other students to comment on the presentation. Occasionally, the teacher may offer a counter-example requiring the student to tighten up his or her argument.

This kind of teaching develops into a culture of mutual respect between teachers and students. And, while it is true that students are competing to become the first to present a proof or a solution, they also become colleagues, but not collaborators in the study of calculus.

A thing that sometimes happens is that a student may wish to obtain his or her own proof and will not want to see the proof that another student is about to present. This gives the teacher an excellent opportunity to respect this situation by allowing such students to leave the room and be called on at a later time to make their own presentation.

I feel the need to admit that my admiration for Dr. Moore is conflicted because of his racism. He did everything he could to block the admission of any student of color into his classes in the 1950s even after the University of Texas in the 1940s had become one of the first institutions in the South to desegregate their halls. African American graduate students who were interested in learning Point-set Topology, which was Dr. Moore s field, were deprived of being able to study with him. But, unfortunately, for him, he missed the chance to engage with some of these same mathematicians who were emerging in the 1960s.

I am sorry that this, otherwise, iconic person, had to have this dark side. Despite this appalling circumstance, I am not ready to abandon his (the IBL) teaching method. In my own 40-year teaching career I have used it in my classrooms.

Many years later, I served in the U. S. Peace Corps in Zimbabwe and was able to teach mathematics using IBL. In the following picture I am returning test papers to students in my Fourth Form in a rural school, near Mutare.

How to Use this Book

I have written this book with the thought that it will be used in IBL calculus classes. In this book the students will encounter a question or problem and then they will be given the opportunity to answer the question or solve the problem before moving on. The solutions given in the text and in the answer section at the back of the text are stated in a step-by-step manner that lets the student uncover the solution one line at time to check his or her work. The student will answer the questions, solve the problem, then check to see if the book gets the same thing, and why or why not?

Ideally, this book would be used in a teacher-facilitated classroom of several students competing for the privilege of presenting their solutions before the class. The organization of the material here maximizes the reader's opportunity to participate in the creative process. Here is how that happens.

- The book is divided into chapters that roughly represent two or three classroom periods.

- Each chapter includes any definition, example, or instruction that facilitates the solution of the problems,

- The questions raised in each chapter are immediately followed by their answers. These are inclosed between two heavy horizontal lines allowing readers to cover up the solution and solve the problem for themselves before reading the answer. An example of this is:

Definition: Subset

The statement that "A is a subset of B " means that A and B are sets and there is no element of A which is not in B.

Problem: Prove that the empty set is a subset of any set

Show that if B is any set, then the empty set $\emptyset$ is a subset of B.
   Solution

Suppose, on the contrary, that there exists a set B such that the empty set $\emptyset$ is not a subset of B, then by definition there must be some element in $\emptyset$ which is not in B. But this is not possible because there are no elements in $\emptyset$. This contradicts the supposition.

Exercises

At the end of each chapter is a set of exercises (the homework problems) to be solved and presented in subsequent periods.

- Detailed step-by-step solutions to the exercises are contained in the final section of this book. Students may use this section to check their work.

- Students may also use the solutions as *Hints*. After trying to work a problem, look at the answer one line at a time. Simply cover all but the first line, and go back and try to work the problem on your own. If this is not sufficient try uncovering the next line and so forth.

- If this is being used as a textbook for a class, the teacher may wish to first, call on a student who has not looked at the answers before calling on one who had. The advantage to the whole class, and to the teacher, is that the student may have discovered a different way to solve the problem.

- Every student in class should have a chance to present his or her solution to the class. An essential part of an IBL class is for the teacher to not always call on the same two or three students for all the solutions.

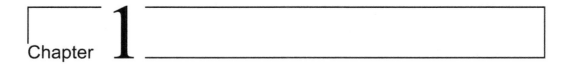

Chapter **1**

Rapid Sketching

1.1 Equations and their graphs

- Importance of graphs.

- Analyzing an equation, to sketch its graph.

- Graphs with positive and negative branches.

1.2 Importance of graphs

During the 300 years preceding Archimedes and Apollonius (about 225 BC), Greek mathematics flourished: Pythagoras in 540 BC, Aristotle in 340 BC, and Euclid in 300 BC. *Euclid's Elements* is a collection of theorems of geometry, organized in such a manner that students of mathematics could prove the theorems and construct various plane and solid geometric figures. This type of geometry is sometimes called *Synthetic Geometry*, meaning geometry that is built up or constructed. In other words, synthesized.

In 1630, Rene Descartes devised a new way to approach geometry   He treated certain geometric entities such as the *conic sections*, (straight lines, circles, ellipses, parabolas, and hyperbolas) by "breaking down" or analyzing them. In addition Descartes introduced the number line and a coordinate system for the two-dimensional plane.   The Cartesian Coordinates were actually anticipated around 150 BC when Hipparchus devised a longitude and

1

latitude system for maps.   The result of Descartes' study is called *Analytic Geometry*.

By using a coordinate system we can assign each point in a geometric figure a pair of real numbers (x, y).   The importance of this is that we can now treat these figures as graphs, with equations that relate x with y. For example, we can say that $y = x$ is the equation of a straight line in the plane.   Also, we can use algebraic information to give us some more geometric information.   For example, $y = x + 1$ is the equation of another straight line in the plane, parallel to the straight line $y = x$.

Graphs are not only useful in geometry, but they are often valuable ways to display information connecting various quantities, such as, distance, time, and speed.   Calculus is simply an extension of analytic geometry   In this course one of the first things we will study is how to get graphs from equations.   Then we will learn how to find maximum and minimum values of one quantity that varies with respect to another quantity, or how rapidly one quantity varies with respect to another, as well as other relationships between variables that are useful in business, science, engineering and mathematics.

1.3   Sketching graphs by analysis

We are going to sketch graphs, not by plotting points, but by analyzing equations.   First, we will find what values substituted for x in an equation will make $y = 0$.

1.3.1   Example: Graph of $y = (x - 1)(x - 2)$.

Sketch the graph whose equation is

$$y = (x - 1)(x - 2). \tag{1.1}$$

1.3.2   Question: Where is y zero?

What values of x will make $y = 0$ in the equation?

Answer

_____

Notice that $x = 1$ or $x = 2$ makes $y = 0$. Therefore, the points $(1, 0)$ and $(2, 0)$ are on the graph. Denote the point $(1, 0)$ by A, and the point $(2, 0)$ by B, and show these points on the coordinate axes. See Figure 1.1

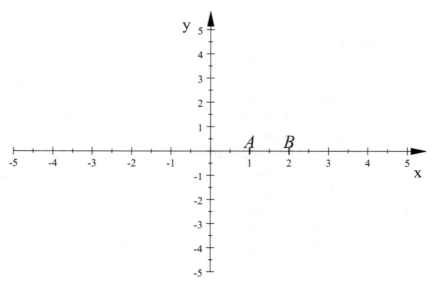

Figure 1.1 The points $A = (1, 0)$ and $B = (2, 0)$.

If we substitute any number for x into the equation, we will get some number for y, and the point (x, y) is a point of the graph. If y is positive then the point (x, y) is above the x-axis, but if y is negative, then the point (x, y) is below the x-axis..

Notice that, in the equation, $y = (x - 1)(x - 2)$, if $x = 0$, then $y = (0 - 1)(0 - 2) = (-1)(-2) = +2$.

1.3.3 Question: If $x < 1$, what sign will y have?

If we substitute *any* number x, less than 1, into the equation $y = (x-1)(x-2)$ what sign ($+$ or $-$) will the number y have?

Answer

When $x < 1$, then it is also true that $x < 2$, so both $x - 1 < 0$ and $x - 2 < 0$, therefore y will be positive because $y = (-)(-) = (+)$.

1.3.4   Question: Where is the graph when $x < 1$?

When $x < 1$, where are all the points of the graph?

Answer

The graph is above the x-axis for all $x < 1$. See Figure 1.2.

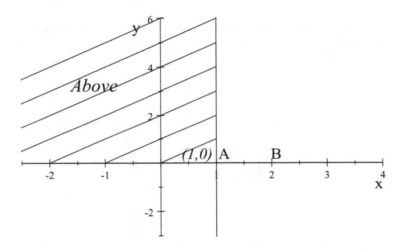

Figure 1.2 $y > 0$ in the region of the plane where $x < 1$.

Now move on to where x is greater than 1, but less than 2, for example if $x = 1\frac{1}{2}$, then $x - 1 = \frac{1}{2}$ and $x - 2 = -\frac{1}{2}$, making $y = (\frac{1}{2})(-\frac{1}{2}) = -\frac{1}{4}$.

1.3.5   Question: If $1 < x < 2$ what sign does y have?

In general, if x is any number such that $1 < x < 2$, what *sign* does y have in Equation (1.1)?

Answer

If x is any number between 1 and 2, then $(x-1) > 0$ and $(x-2) < 0$, and since $y = (x-1)(x-2)$, then $y = (+)(-) = (-)$ so $y < 0$ whenever $1 < x < 2$.

1.3.6   Question: If $1 < x < 2$ where is the graph?

Where are all the points of the graph relative to the x-axis, when $1 < x < 2$?

Answer

The graph is below the x-axis because $y < 0$ when $1 < x < 2$. See Figure 1.3

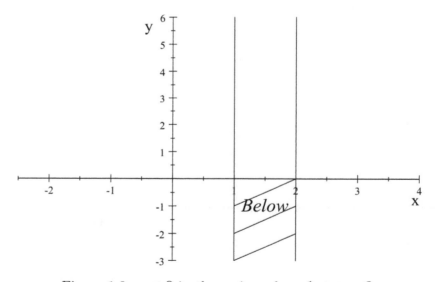

Figure 1.3 $y < 0$ in the region where $1 < x < 2$.

Finally, when x is greater than 2, we have that x is greater than 1 as well. Hence, each of $(x-1)$ and $(x-2)$ is a positive number. And since y is the product of these two positive numbers, then y must be positive, and the graph G is above the x-axis for all x beyond 2. See Figure 1.4

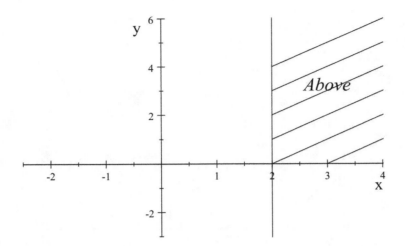

Figure 1.4. Region where $x > 2$ and $y > 0$.

In the next graph we combine the regions found in Figures 1.1, 1.2, 1.3, and 1.4.

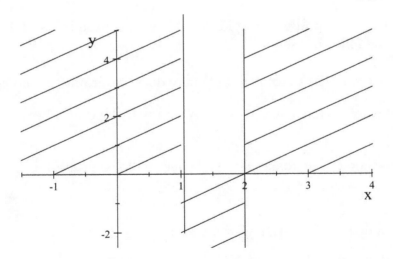

Figure 1.5. Combined regions.

If we plot points from the shaded region above the x-axis to the left of $x = 1$, into the shaded region below the x-axis between 1 and 2 and then into the shaded region above the x-axis to the right of $x = 2$, we get the graph in Figure 1.6.

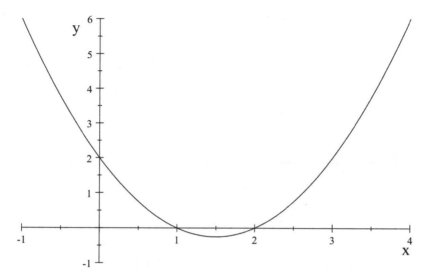

Figure 1.6 Graph of $y = (x - 1)(x - 2)$.

1.3.7 Summary:

Given an equation in x and y, a "rapid sketch" of the graph is a sketch based upon finding the following sets:

- The values of x that make $y = 0$. Where does the graph cut the x-axis?

- The values of x that make $y > 0$. Where will the graph be above the x-axis?

- The values of x that make $y < 0$. Where will the graph be below the x-axis?

1.3.8 Problem: Graph $y = (x - 1)(x - 2)(x - 3)$

Obtain a rapid sketch of the graph whose equation is:

$$y = (x - 1)(x - 2)(x - 3). \qquad (1.2)$$

Solution

- When $x = 1$, $x = 2$, or $x = 3$, then $y = 0$; so the graph cuts the x-axis at the points $(1, 0), (2, 0)$ and $(3, 0)$.

- When $x < 1$, then $y < 0$; so the graph is below the x-axis.

- When $1 < x < 2$, then $y > 0$; so the graph is above the x-axis.

- When $2 < x < 3$, then $y < 0$; so the graph is below the x-axis.

- When $x > 3$, then $y > 0$; so the graph is above the x-axis.

In Figure 1.7, we show these shaded regions, for Equation (1.2), then in Figure 1.8 we show the graph as it passes from one region to another.

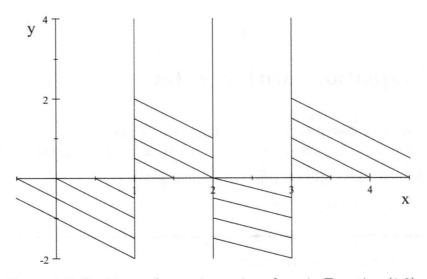

Figure 1.7. Positive and negative regions for y in Equation (1.2).

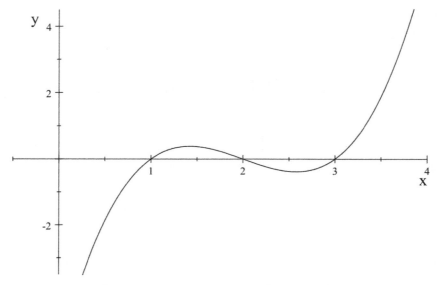

Figure 1.8 Graph of $y = (x-1)(x-2)(x-3)$.

1.4 Equations with a y^2 term

We are going to sketch two graphs, G and H. The equation for G will have the variable y to the first power; in the equation for H we replace y by y^2.

(a) Sketch the graph G, whose equation is $y = x$.

(b) Sketch the graph H, whose equation is $y^2 = x$.

(a) See Figure 1.9 for the graph of $y = x$.

(b) See Figure 1,10 for the graph of $y^2 = x$.

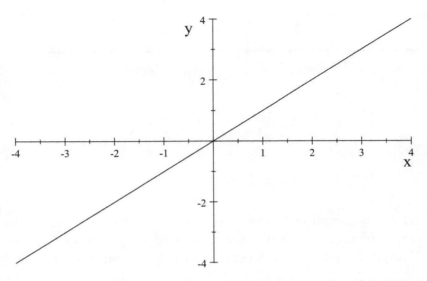

Figure 1.9 Graph of $y = x$.

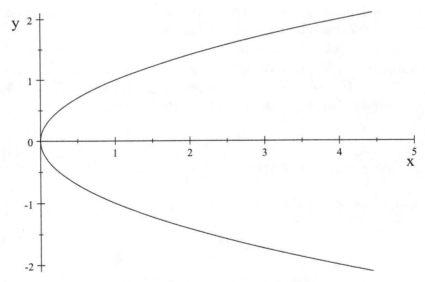

Figure 1.10 Graph of $y^2 = x$.

1.4.1   Question: How did we get the graph of $y^2 = x$?

Why is H the graph of the equation $y^2 = x$?

Answer

This one single equation, $y^2 = x$, is equivalent to the following set of two equations:

$$\begin{aligned} y &= \sqrt{x} \\ y &= -\sqrt{x}. \end{aligned}$$

Therefore the graph H has two branches as shown in Figure 1.10. Think of the graph H as being two square roots of G, wherever G is above the x-axis. But H is nonexistent whenever G is below the x-axis.

Every positive number has two square roots, but a negative number does not have a (real) square root. For example, the number 4 has two square roots, $\pm \sqrt{4} = \pm 2$, but the number -1 has no square root.

When we re-write the equation $y^2 = x$ as $y = \sqrt{x}$ and $y = -\sqrt{x}$ we are assuming that x is positive, we cannot take the square root of x if x is negative. Thus, H, the $\pm$ square root graph of G, can only exist for those values of x for which the graph G is above the x-axis.

1.4.2 Problem: Graph $y^2 = (x-1)(x-2)$

Obtain a rapid sketch of the graph whose equation is:

$$y^2 = (x-1)(x-2). \tag{1.3}$$

Solution

First, let's go back to the equation with y to the first power, not the y^2 equation. That is, go back to the graph whose equation is $y = (x-1)(x-2)$. This was sketched in Figure 1.6, which we repeat here. Call this graph K.

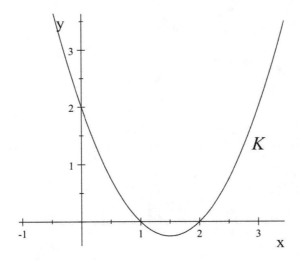

Figure1.11 Repeat of Graph of $y = (x-1)(x-2)$.

Notice that K is above the x-axis when $x > 2$ or $x < 1$, but it is below the x-axis when x is between $x = 1$ and $x = 2$.

Now, if we let L be graph of Equation (1.3), then we are really sketching the graph of two equations that are equivalent to Equation (1.3), namely

$$
\begin{aligned}
y &= \sqrt{(x-1)(x-2)} \\
y &= -\sqrt{(x-1)(x-2)}.
\end{aligned}
$$

In the graph L, shown in Figure 1.12, we sketch the graphs of these two equations by drawing two branches , the $\pm$ square roots of K. The graph L only exists when K is above the x-axis.

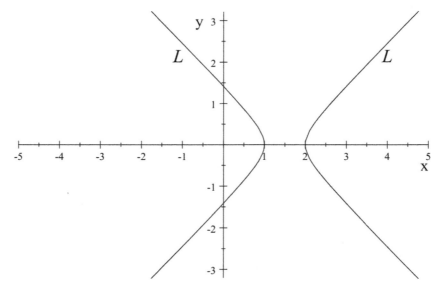

Figure 1.12 Graph of $y^2 = (x - 1)(x - 2)$.

1.4.3 Problem: Graph $y^2 = (x - 1)(x - 2)(x - 3)$

Sketch the graph of

$$y^2 = (x - 1)(x - 2)(x - 3). \tag{1.4}$$

Solution

Equation (1.4) may be written as the two equations

$$y = \sqrt{(x - 1)(x - 2)(x - 3)} \tag{1.5}$$

$$y = -\sqrt{(x - 1)(x - 2)(x - 3)}. \tag{1.6}$$

These equations are square roots of Equation (1.2), which is: $y = (x - 1)(x - 2)(x - 3)$ whose graph we sketched in Figure 1.8.

Let us take another look at this graph in Figure 1.13.

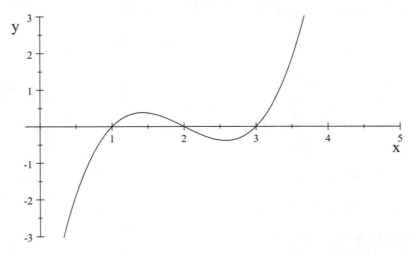

Figure 1.13 Repeat of the Graph of $y = (x-1)(x-2)(x-3)$.

Notice that the graph of $y = (x-1)(x-2)(x-3)$ is above the x-axis only for x between 1 and 2 or for all $x > 3$, and it is below the x-axis when x is less than 1 or x is any number between 2 and 3.

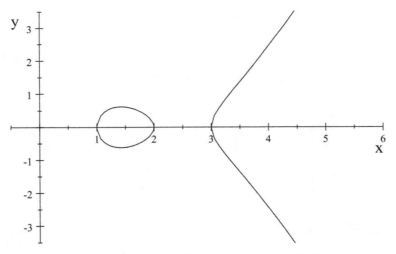

Figure 1.14. Graph of $y^2 = (x-1)(x-2)(x-3)$.

Therefore two square roots ($\pm$) of the graph exists for all x between 1 and 2, or for all $x > 3$, but a square root does not exist for any x less than 1, nor for any x between 2 and 3. See Figure 1.14.

1.5 Exercise 1 HOMEWORK

For each of the following equations, rapid sketch the graph.

1. $y = -(x-1)(x-2)$.

2. $y^2 = -(x-1)(x-2)$.

3. $y = (x-1)^2(x-2)$.

4. $y^2 = (x-1)^2(x-2)$.

5. $y = x(x+2)(x-3)$.

6. $y^2 = x(x+2)(x-3)$.

Chapter **2**

Lowest Point of a Graph

2.1   The importance of maxima and minima

When we draw a graph by rapid sketching, we learn what is happening to y overall. That is, we answer the questions: Where is y positive? Where is $y = 0$? And where is $y < 0$? In other words, we can see where the graph is above the x-axis, where it crosses the axis and where it's below the axis. But this analysis is not fine enough to determine how high the graph goes before it comes back down, or how low it gets before it starts back up. These are important places on the graph. They are called "maximum points" or "minimum points" of the graph. Usually, they are not absolute maxima or minima, because they may just be high points or low points over a short piece of the graph. Perhaps the graph continues up forever as x increases; so there is no absolute maximum. But when the graph has a point that is higher than all the points for a short piece on either side of it, then we call such a point a "relative maximum" as opposed to an "absolute maximum".

The reason we should care about what point is a (relative) maximum or minimum of graph is that such a point could represent the best or the worst you could do in some application. For example, the graph could be one in which y is the velocity of a rocket based upon the length of time, x, that the rocket fuel burns. You will very likely wish to determine the relative maximum at the exact moment of launch. If the burn time is too short (or too long) the velocity may not be at its highest point.

In calculus, there is a method for finding what the relative maxima or relative minima of a graph are. We will show how we can use the slope of a graph to find the point at which the graph "tops-out" or "bottoms-out".

Before going any further, solve the following problems.

2.1.1   Problem: A little bit of Factoring

Factor the difference of two cubes $a^3 - b^3$.

Solution

$$a^3 - b^3 = (a - b)(a^2 + ab + b^2).$$

2.1.2   Problem: Rationalizing the *numerator*

Rationalize the numerator in the fraction, and simplify

$$\frac{\sqrt{r} - \sqrt{s}}{r - s}.$$

Solution

Multiply the numerator and denominator by $\sqrt{r} + \sqrt{s}$, getting

$$\frac{(\sqrt{r}-\sqrt{s})(\sqrt{r}+\sqrt{s})}{(r-s)(\sqrt{r}-\sqrt{s})} = \frac{(r-s)}{(r-s)(\sqrt{r}+\sqrt{s})} = \frac{1}{\sqrt{r}+\sqrt{s}}.$$

2.2   Finding the lowest point of a graph

We want to find the lowest point of a graph between two given numbers.

Example: Given the equation:

$$y = (x - 1)(x - 2)(x - 3). \tag{2.1}$$

Find the lowest point of its graph between $x = 2$ and $x = 3$. The graph of this equation is shown in Figure 2.1. Also shown is the line PQ cutting the graph in two points P and Q. Such a line is called a *secant* line for the graph, (The word *secant* means *cutting*.) The coordinates of the points P and Q are determined by Equation (2.1). Thus, if p is the abscissa of the point P, then its ordinate is $(p-1)(p-2)(p-3)$.

Notice that this graph does not have a lowest point over the whole x-axis, but it does have a lowest point in a small neighborhood, such as the set of all x between $x = 2$ and $x = 3$.

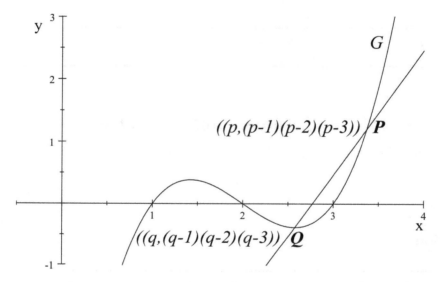

Figure 2.1 Graph of Eq. (2.1) and the line PQ.

In Figure 2.1, we denote by Q the lowest point of G between $x = 2$ and $x = 3$. Let q be the abscissa of Q. We don't yet know what value q has. Let us suppose that the other point P on G is allowed to slide along the graph toward Q.

No matter what location P has on the graph (as long as it is not yet Q), we say that its abscissa is p and its ordinate is $(p-1)(p-2)(p-3)$. Thus, the coordinates of P and Q are:

$$\begin{aligned} Q &= (q, (q-1)(q-2)(q-3)) \\ P &= (p, (p-1)(p-2)(p-3)). \end{aligned}$$

By definition, if P and Q are any two points, the slope of the line PQ is the difference in their ordinates divided by the difference in their abscissas.

Denote by m_{PQ} the slope of the secant line PQ, then

$$m_{PQ} = \frac{(p-1)(p-2)(p-3) - (q-1)(q-2)(q-3)}{p-q}. \qquad (2.2)$$

Multiply out the factors

$$m_{PQ} = \frac{p^3 - 6p^2 + 11p - 6 - q^3 + 6q^2 - 11q + 6}{p-q}.$$

Cancel the -6 and the $+6$, and re-group the other terms

$$m_{PQ} = \frac{p^3 - q^3 - 6(p^2 - q^2) + 11(p-q)}{p-q}.$$

Factor:

$$m_{PQ} = \frac{(p-q)(p^2 + pq + q^2) - 6(p-q)(p+q) + 11(p-q)}{(p-q)}.$$

Simplify

$$m_{PQ} = p^2 + pq + q^2 - 6p - 6q + 11. \qquad (2.3)$$

In Equation (2.3), m_{PQ} is the slope of the line through the lowest point Q, and any point P on G. It is the best way to write the slope of the line PQ because it no longer has the factor $(p-q)$ in the denominator.

2.2.1   Question: Why cancel $(p-q)$ from the denominator?

Why is it better to have the factor $(p-q)$ cancelled out of the denominator?

Answer

Because we can let the number p approach the number q without having the denominator approach zero. That is, we have eliminated the possibility of dividing by zero.

If we let the point P approach the point Q along the graph then we get a line that is tangent to G at the point Q. The tangent line touches the graph at the point Q, but does not cross over to the other side of the graph at Q. The word *tangent* means *touching*.

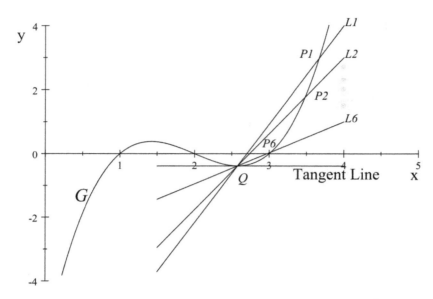

Figure 2.2. Lines approaching the tangent line at Q.

In Figure 2.2, the points $P1$, $P2$, ..., $P6$, ... are supposed to represent, geometrically, that the point P is moving along the curve towards the point Q, that is: $P \to Q$, the line PQ becomes the tangent line at Q.   But, in addition to this (geometric) movement in which one point approaches another, there is also a numerical change taking place! The abscissa p of P is approaching the abscissa q of the point Q. This change affects the slope, m_{PQ} given in the Equation (2.3).

Question: What happens to $p^2 + pq + q^2 - 6p - 6q + 11$ as $p \to q$?

We break this one question down into six smaller questions, as in the following call and response litany.

"So what happens to p^2, when $p \longrightarrow q$?"

Class: "It approaches q^2".

"What happens to pq when $p \longrightarrow q$?"

Class: "It approaches q^2".

"What happens to q^2, when $p \longrightarrow q$?"

Class: "It stays q^2".

"Now what happens to $-6p$ when $p \longrightarrow q$?"

Class: "It approaches $-6q$".

"What happens to $-6q$ when $p \longrightarrow q$?"

Class:"It stays $-6q$".

"What happens to 11 as $p \longrightarrow q$?"

Class: "It stays 11".

Thus the line tangent to G at Q will have slope $m = q^2 + q^2 + q^2 - 6q - 6q + 11$. That is, the slope of the tangent line to G at the point Q is:

$$m = 3q^2 - 12q + 11. \tag{2.4}$$

2.2.2   Question: What is the tangent at a low point?

Using your intuition, answer the following question: If Q is the lowest point of a graph, what can be said about the line tangent to the graph at Q?

Answer

A tangent line at the lowest point of a graph must be *horizontal*. Its slope must be zero; thus, $m = 0$.

2.2.3   Question: Where is the slope $= 0$?

What value of q makes $m = 0$ in Equation (2.4)?

Answer

If $3q^2 - 12q + 11 = 0$, then by the quadratic formula:

$$q = \frac{12 \pm \sqrt{12^2 - 4 \times 3 \times 11}}{2 \times 3} = 2 \pm \frac{\sqrt{3}}{3}.$$

Denote these two roots by q_1 and q_2. So at either $q_1 = 2 + \frac{\sqrt{3}}{3}$ or $q_2 = 2 - \frac{\sqrt{3}}{3}$ G will have a tangent line with slope, $m = 0$.

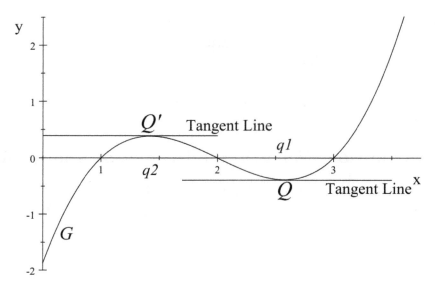

Figure 2.3. Q is a relative minimum and Q' is a relative maximum.

From the graph in Figure 2.3, we see that q_1 is the abscissa of the lowest point of G between $x = 2$ and $x = 3$, and q_2 is the abscissa of the highest point of G between $x = 1$, and $x = 2$. Since $(q_1 - 1)(q_1 - 2)(q_1 - 3) = \frac{-2\sqrt{3}}{9}$, then the coordinates of Q are $(2 + \frac{\sqrt{3}}{3}, \frac{-2\sqrt{3}}{9})$.

Let Q' denote the highest point of the graph between $x = 1$ and $x = 2$, then the coordinates of Q' are $(2 - \frac{\sqrt{3}}{3}, \frac{2\sqrt{3}}{9})$. In Figure 2.3, we show the points Q and Q' and the lines tangent to the graph at these points.

2.3   Exercise 2 HOMEWORK

1. Let G be the graph of $y = (x - 1)^2(x - 2)$. Sketch G and find the coordinates of its lowest point between $x = 1$ and $x = 2$.

2. Let G be the graph whose equation is $y = x(x + 2)(x - 3)$.

   (a) Rapid sketch the graph G.

   (b) Find the coordinates of the lowest point of G between $x = 0$ and $x = 3$.

3. If G is the graph of $y = (x - 1)(x - 2)$, find the coordinates of a point Q, such that the tangent line to G has slope $m = -1$ at Q.

4. Given the equation

$$p = \frac{\sqrt{r^2 + r + 2} - \sqrt{s^2 + s + 2}}{r - s}$$

   (a) "Rationalize" the numerator and simplify so that the factor $(r - s)$ will be cancelled out of both the numerator and denominator.

   (b) After simplifying as required in part (a), let $r = s$ and find p.

5. Let G be the graph of the equation $y^2 = x(x + 2)(x - 3)$.

   (a) Rapid sketch G.

   (b) Find the coordinates of the lowest and highest points of this graph between $x = -2$ and $x = 0$.

Chapter 3

Asymptotes

3.1 Asymptotes

- Getting closer and closer without ever reaching.

- We will see what happens to an expression when a variable increases
  or decrease indefinitely.

3.2 Getting closer without reaching

When a "bungee jumper" hurls himself off a high place, he goes plummeting
down toward the ground at a rapid speed, but when he reaches the length
of the chord, it starts to slow him down, we hope, and, if he is lucky, his
rate of falling will slow down more and more as the chord stretches and he
gets closer and closer to the ground. Eventually, he will slow down to zero
velocity at a short distance above the ground, we hope. This gradual "closing-
in" on ground level can be described as "approaching without reaching"
zero. Ideally, if time could go on forever, while his distance from the ground
decreased by half each second, then he could fall for ever and get ever closer
and closer to the ground.

This is an example of one variable approaching a fixed number, like height
approaching zero, while the other variable continues indefinitely, like the time
it takes to fall going on forever.

3.2.1   Question: What does $7 + \frac{10}{x}$ approach?

If x increases without bound, what happens to the expression $7 + \frac{10}{x}$?

Answer

$7 + \frac{10}{x}$ approaches $7 + 0$, or 7, as $x \to \infty$, because the fraction $\frac{10}{x}$ decreases in absolute value, and actually approaches 0 as x increases without bound. For example, if $x = 10$, $\frac{10}{x} = 1.0$, if $x = 100$, then $\frac{10}{x} = 0.1$, if $x = 1000$, then $\frac{10}{x} = 0.01$, etc.

3.2.2   Definition: FRACTION

The number $\frac{a}{b}$ is defined to be the only number that can be multiplied by b to give you a.

3.2.3   Question: What is $\frac{1}{0}$?

Why is $\frac{1}{0}$ is not defined?

Answer

If $\frac{1}{0}$ were a number, it would have to be the only number that when multiplied by 0 would give you 1. But any number multiplied by 0 gives you 0 not 1.

3.2.4   Question: What happens to $\frac{10}{x}$ as $x \to 0$?

What happens to the fraction $\frac{10}{x}$ where $x \neq 0$, but is approaching zero?

Answer

The smaller the denominator, the more "times" it will go into the numerator; so the fraction will be larger in absolute value. For example if

$x = 1$, $\frac{10}{x} = 10$, but if $x = 0.1$, $\frac{10}{x} = 100$, if $x = 0.01$, $\frac{10}{x} = 1000$, etc. Thus, when $x > 0$, the fraction $\frac{10}{x}$ increases indefinitely as $x \to 0$. In other words, $\frac{10}{x} \to \infty$, when $x \to 0$.

3.2.5 Question: What happens to $\frac{x-1}{x-2}$ as $x \to 2$?

What happens to the fraction $\frac{x-1}{x-2}$, where $x \neq 2$, but x is approaching 2?

Answer

If $1 < x < 2$ and x is approaching 2, the numerator is positive and approaching $+1$, but the denominator is negative and approaching 0. So the fraction $\frac{x-1}{x-2}$ is negative and increasing indefinitely in absolute value, so we say $\frac{x-1}{x-2} \to -\infty$, as $x \to 2$ from the left. Similarly if $x > 2$ and $x \longrightarrow 2$, the fraction gets indefinitely large; thus, $\frac{x-1}{x-2} \longrightarrow +\infty$.

3.2.6 Problem: Graph of $y = \frac{x-1}{x-2}$.

Rapid sketch the graph G whose equation is

$$y = \frac{x-1}{x-2}. \tag{3.1}$$

Solution

For this equation we make the following observations:

1. If $x = 1$, then $y = 0$.

2. If $x = 2$, then the Equation (3.1) would become $y = \frac{2-1}{2-2}$, or $\frac{1}{0}$ which, as we just saw, is not defined.

3. When $x < 1$, the graph is above the x-axis because then both $x - 1$ and $x - 2$ are negative, so $y > 0$.

4. Also when $x > 2$, the graph is above the x-axis because then both $x - 1$ and $x - 2$ are positive, so $y > 0$ again.

5. The only values of x that make $y < 0$ are between $x = 1$ and $x = 2$; that is when $1 < x < 2$, because then $x - 1 < 0$ and $x - 2 > 0$, making $\frac{x-1}{x-2} < 0$, and putting the graph below the x-axis.

6. When the absolute value of x increases indefinitely, that is, when $x \to \infty$, or when $x \to -\infty$, we shall see that y approaches 1. This means that the graph will approach the horizontal line $y = 1$.

7. We will also see that $y \to \infty$, or $y \to -\infty$ when $x \to 2$.

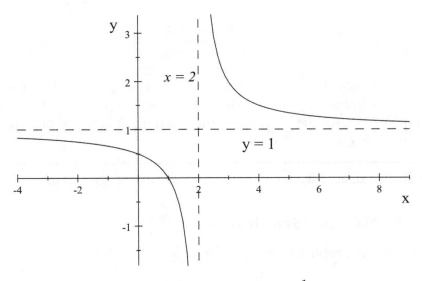

Figure 3.1 Graph of $y = \frac{x-1}{x-2}$.

The two dashed lines $x = 2$ and $y = 1$ in Figure 3.1 are not a part of the graph of the equation, but are *asymptotes*, as defined below.

3.2.7   Definition: ASYMPTOTE

When a graph approaches a line, getting closer and closer to that line as one of the variables x or y increases or decreases indefinitely, then that line is called an asymptote of the graph.

In Figure 3.1 the graph approaches the vertical line $x = 2$; so this line is a vertical asymptote for the graph making $y \to +\infty$ as $x \longrightarrow 2$ when $x > 2$, and $y \longrightarrow -\infty$ as $x \to 2$ when $x < 2$.

3.2.8 Question: What happens to $\frac{x-1}{x-2}$ as $x \to \infty$?

In the equation $y = \frac{x-1}{x-2}$, what happens to y when x increases indefinitely?

Answer

We can divide the numerator and denominator of $\frac{x-1}{x-2}$ by x getting:

$$y = \frac{1 - \frac{1}{x}}{1 - \frac{2}{x}}.$$

Now as $x \to \infty$, $\left(1 - \frac{1}{x}\right) \to 1$, and $\left(1 - \frac{2}{x}\right) \to 1$, so $y \to \frac{1}{1}$. This means that as $x \to \infty$, the graph approaches the horizontal line $y = 1$, and as $x \to -\infty$, the graph also approaches this same line, making it a horizontal asymptote of the graph. See Figure 3.1.

Now let us again turn to an equation containing a y^2 term.

3.2.9 Problem: Graph of $y^2 = \frac{x-1}{x-2}$

Rapid sketch the graph whose equation is

$$y^2 = \frac{x-1}{x-2}. \tag{3.2}$$

Solution

This equation may be written as

$$y = \sqrt{\frac{x-1}{x-2}} \tag{3.3}$$

$$y = -\sqrt{\frac{x-1}{x-2}}. \tag{3.4}$$

This means we need only look at the graph in Figure 3.1, and sketch two branches $\sqrt{\frac{x-1}{x-2}}$, and $-\sqrt{\frac{x-1}{x-2}}$ only when $\frac{x-1}{x-2} > 0$. See Figure 3.2.

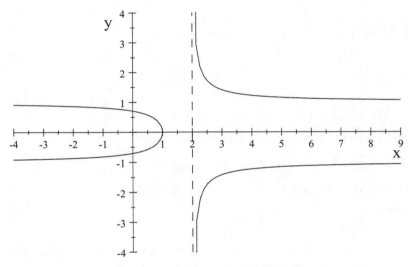

Figure 3.2 Graph of $y^2 = \frac{x-1}{x-2}$.

3.3   Exercise 3 HOMEWORK

1. (a) Sketch the graph G whose equation is

$$y = \frac{(x-1)(x-2)}{(x-3)}.$$

   (b) What happens to the graph for $x > 3$ as x increases indefinitely? That is, what happens as $x \longrightarrow \infty$?

   (c) Find the coordinates of the highest point of G between $x = 1$ and $x = 2$.

2. Rapid Sketch the graph H, whose equation is

$$y^2 = \frac{(x-1)(x-2)}{(x-3)}.$$

3. Rapid Sketch the graph whose equation is

$$y^2 = \frac{(x-1)(x-2)^2(x-3)}{(x-4)}.$$

4. Let K be the graph of the equation

$$y = \frac{x+1}{x^2+x+1}.$$

   (a) Find every point where K crosses the x-axis.

   (b) Find the highest and the lowest points of K.

   (c) What happens to K as $x \to \infty$?

   (d) What happens to K as $x \to -\infty$?

   (e) Sketch the graph of K.

5. Let L be the horizontal line whose equation is $y = \frac{1}{10000}$, and let G be the graph whose equation is $y = \frac{1}{x}$.

   (a) Show that there is some positive number p, such that the point $P = (p, \frac{1}{p})$ of G is between L and the x-axis.

   (b) If p is one of the numbers found above in Part (a), show that for all numbers $t > p$, the points $(t, \frac{1}{t})$ are also between L and the x-axis.

   (c) Is there a smallest positive number p such that the point $(p, \frac{1}{p})$ of G is between L and the x-axis?

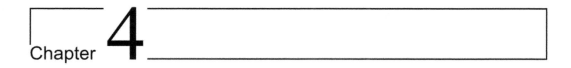

Chapter 4

Simple Graphs

4.1 Questions about graphs and slope

- What is a simple graph?

- What is *function notation*?

- What is the slope of a graph?

- What can we learn from the slope?

- When does a graph have zero slope?

- When does a graph have no slope?

4.2 Simple graphs

4.2.1 Definition: SIMPLE GRAPH

A simple graph is a set of one or more points in the coordinate plane, such that no vertical line contains two points of it. This means that it is not possible to draw a vertical line that intersects it twice. See Figure 4.1 for an example of a simple graph.

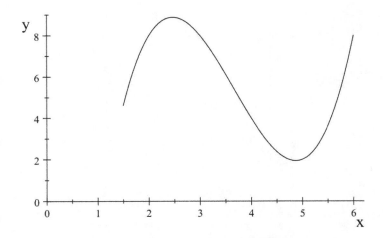

Figure 4.1 An example of a Simple Graph.

4.2.2   Question: Some Logic

If you are told, "No one in our city has two credit cards". Does this mean that *every* person in our city has, at least, one credit card? Is it possible that someone in our city might have three or more credit cards?

Answer

No, it does not mean that every person in our city has least one credit card. No, it is not possible that someone might have two or more credit cards, because such a person would then have at least two. The original statement is true only if each person has either *no credit card* or *exactly one*.

4.2.3   Question: Is the graph in Figure 4.2 a simple graph?

Look at the graph in Figure 4.2. Is it a simple graph?

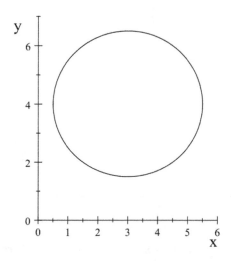

Figure 4.2 Is this a simple graph?

The graph in Figure 4.2 is *not* a simple graph because it is possible to draw a vertical line that contains two points of it. For example, if you draw the vertical line with equation $x = 3$, it will cut the graph twice.

4.2.4   Definition: FUNCTION NOTATION

If G is a simple graph and the point (x, y) is a point of G, then, in order to identify y as the ordinate of the point whose abscissa is x, we may denote y as $f(x)$, where $f(x)$ is some algebraic expression in x. Thus, every point of G is represented by $(x, f(x))$. The expression $f(x)$ is called *function notation*. The equation of G is

$$y = f(x). \tag{4.1}$$

4.2.5   Example: What is $f(-1)$ when $f(x) = 2x^3 + x^2 - 10x - 3$?

If $f(x) = 2x^3 + x^2 - 10x - 3$, and G is the graph whose equation is $y = f(x)$, what is the ordinate of the point A if the abscissa of A is -1?

Solution

Since the abscissa of A is -1 then the ordinate of A is $f(-1)$, which is $2(-1)^3 + (-1)^2 - 10(-1) - 3 = 6$.

The function notation $f(x)$ is a kind of "shorthand" that we may use for any expression in x.

4.2.6   Example: If $f(x) = \frac{x-3}{x^2+3}$ what is $f(-3)$?

If $f(x) = \frac{x-3}{x^2+3}$ and G is the graph whose points are $(x, f(x))$, find the ordinate of a point P whose abscissa is -3.

Solution

Here when $x = -3$, the function value $f(-3)$ is $\frac{(-3)-3}{(-3)^2+3} = \frac{-6}{12} = -\frac{1}{2}$.

We can use letters other than $f(x)$ in function notation, such as $g(x)$, or $h(x)$ in place of $f(x)$, as long as we provide the an expression in terms of x. Also the values of x must be such that the expression has meaning. For example if $g(x) = \frac{1}{x-1}$, we can find $g(3)$, but $g(1)$ has no meaning. Why?

We also may use other letters for the variable x. For example, Let H be the graph of whose equation is $y = u(t)$, where $u(t) = t^3 + \sqrt{t+13} + 61$. To find what $u(t)$ would be for a given number t, we simply substitute the given number in for t. Thus, if we wanted to know what $u(t)$ would be if $t = 0$, we have $u(0) = 0^3 + \sqrt{0+13} + 61$. Or approximately $u(0) \approx 64.606$.

4.2.7   Question: What is $u(-4)$?

For the function $u(t) = t^3 + \sqrt{t+13} + 61$, what is $u(-4)$?

Answer

If $t = -4$, $u(t)$ would become $u(-4) = (-4)^3 + \sqrt{-4+13} + 61$. That is, $u(-4) = 0$.

4.2.8 Definition: SLOPE OF A GRAPH

If G is a simple graph and A is a point of G, then the statement that *The number m is the slope of G at the point A*, means that m is the slope of the line *tangent* to G at A.

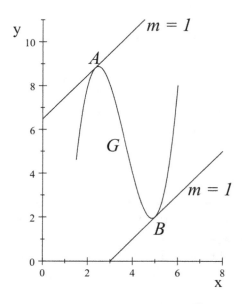

Figure 4.3 These two tangent lines to G have slope 1.

4.2.9 Example: Where does the graph in Figure 4.3 have slope = 1?

The graph G in Figure 4.3 has slope 1 at the points A and B, because each tangent line has a slope of $m = 1$.

4.3 What slope tells you

The slope of a graph at given points tells you something about the *steepness* of the graph at those points. If the slope is positive, it tells you that the graph is rising at that point, and the larger the slope, the faster the graph is rising as you go to the right. See Figure 4.4(a) for a graph that has a positive slope at each of its points. If the slope of the graph is negative at any point, the graph is decreasing at that point, and the more negative, the steeper the descent. See Figure 4.4(b) for a graph that has a negative slope at each of its points.

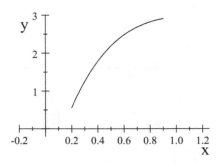 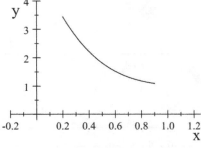

Figure 4.4 (a) A curve with   Figure 4.4 (b) A curve with
positive slope.   negative slope.

If there is some point where the slope of a graph is zero, that is, the tangent line is horizontal, then the graph is neither rising nor falling at that point. In such a case the graph has either reached a maximum point or a minimum point or an inflection point.

4.3.1   Definition: RELATIVE MAXIMUM

A relative maximum is a point where the graph had been previously rising, leveled off then started falling.

4.3.2   Definition: RELATIVE MINIMUM

A relative minimum point is one where the graph had been previously falling, leveled off then started rising.

4.3.3   Definition: INFLECTION POINT

An inflection point is one where the graph levelled off from previously rising, then started rising again, or leveled off from previously falling then started falling again. In Figure 4.5 the graph has slope zero at the three points: A = relative maximum, B =relative minimum and C = inflection point.

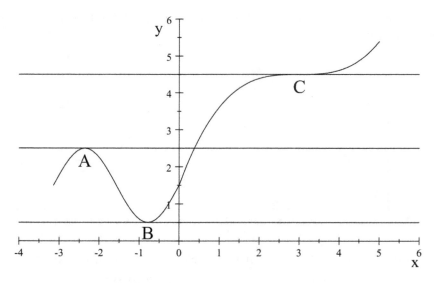

Figure 4.5 Graph has slope zero at A, B and C.

4.3.4 Definition: NO SLOPE

If a graph has a *vertical* tangent at a point, we say that the graph has *no slope* at that point. "No slope" does not mean the same thing as "zero slope". *No slope* means that there is no number that can be called the slope. See the graph in Figure 4.6 where the vertical line $x = 2$ is tangent to the graph at the point $A = (2, 1)$.

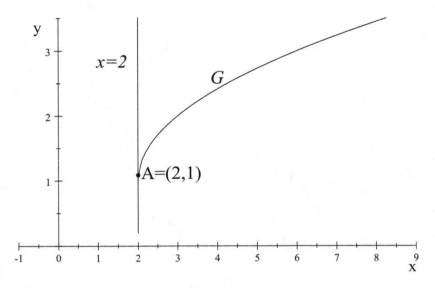

Figure 4.6 The vertical tangent at A means G has no slope at A.

4.4   How to get the slope of a graph

Let G be a simple graph whose equation is $y = g(x)$. Let P and A be two points of G with abscissas p and a respectively, then the coordinates of P and A are $(p, g(p))$ and $(a, g(a))$. See Figure 4.7.

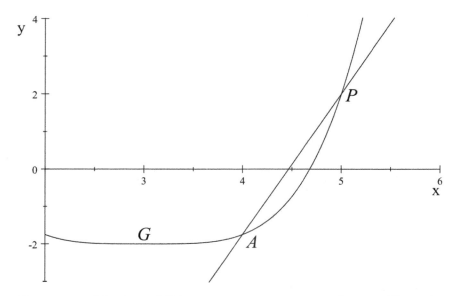

Figure 4.7 The line AP becomes the tangent at A when $P \to A$.

Assume that G does not have a vertical asymptote between A and P. The slope of the line AP will be

$$m_{AP} = \frac{g(p) - g(a)}{p - a},\qquad(4.2)$$

and when we let $P \to A$, we are letting $p \to a$; in other words, the slope of G at A is the number m defined in the following limit,

$$m = \lim_{p \to a} \frac{g(p) - g(a)}{p - a}.$$

If this number m exists, we denote it by $g'(a)$. That is,

$$g'(a) = \lim_{p \to a} \frac{g(p) - g(a)}{p - a}.\qquad(4.3)$$

The main algebraic problem in obtaining the limit in Equation (4.3) is that of determining how to get rid of the factor $p - a$ in the denominator. We need to do this in order to be able to actually let P become A. Thus, we must find a way to let $p \to a$ without producing a 0 in the denominator.

4.4.1   Example: What is the slope of $(3x-5)^2+8$ at $x=a$?

If F is a simple graph whose equation is $y=f(x)$ where

$$f(x) = (3x-5)^2 + 8,$$

find the slope $f'(a)$.

Solution

Let P and A be two points of F with abscissas p and a respectively then $P = (p, f(p))$ and $A = (a, f(a))$. So the slope of PA is

$$
\begin{aligned}
m_{AP} &= \frac{f(p)-f(a)}{p-a} \\
&= \frac{(3p-5)^2 + 8 - ((3a-5)^2 + 8)}{p-a}, \\
&= \frac{(3p-5)^2 - (3a-5)^2}{p-a}.
\end{aligned}
\tag{4.4}
$$

We can factor the numerator as the difference of two squares.

$$(3p-5)^2 - (3a-5)^2 = \left[(3p-5)-(3a-5)\right]\left[(3p-5)+(3a-5)\right],$$

$$
\begin{aligned}
&= \left[3p-3a\right]\left[3p+3a-10\right] \tag{4.5}\\
&= 3\left[p-a\right]\left[3(p+a)-10\right]. \tag{4.6}
\end{aligned}
$$

So, now Equation (4.4) becomes

$$m_{AP} = \frac{3\left[p-a\right]\left[3(p+a)-10\right]}{p-a}.$$

Cancel the factor $(p-a)$ from numerator and denominator and find the slope as $p \to a$ getting

$$\lim_{p \to a} \frac{3\left[3(p+a)-10\right]}{1} = 18a - 30.$$

Thus,

$$
\begin{aligned}
f(x) &= (3x-5)^2 + 8 \text{ implies} \\
f'(a) &= 18a - 30.
\end{aligned}
$$

4.5   Exercise 4 HOMEWORK

1. If a simple graph G has the equation $y = g(x)$, where $g(x) = 2x^4 + K$, and K is any constant, find the slope of G at the point $A = (a, g(a))$.

2. A simple graph F has the equation $y = f(x)$, where $f(x) = x^{1/2}$.

   (a) Find $f'(x)$ at any point $A = (x, f(x))$, where $x \neq 0$.

   (b) Does the graph F have slope at the point $(0, 0)$?

3. If H is a simple graph whose equation is $y = h(x)$, where $h(x) = g(x) + f(x)$ using the functions, $f(x)$ and $g(x)$ from Problems 1 and 2, find $h'(a)$.

4. If C is a simple graph whose equation is $y = c(x)$, where $c(x) = 3h(x)$, using the function $h(x)$ from Problem 3, find $c'(a)$.

5. Let $r(x) = x(x-1)(x+1)$, and R be the graph whose equation is $y = r(x)$.

   (a) Rapid sketch the graph R.

   (b) Find $r'(x)$ at any point $(x, r(x))$.

   (c) Find the coordinates of the points at which R has a slope of 5.

6. A graph has the equation $s(x) = x^3 + x^2 + x + 1$.

   (a) Find $s'(x)$.

   (b) Is there any point where $s'(x) = 0$?

7. Given the function $u(t) = -\frac{1}{2}at^2 + v_0 t + h$, where a, v_0, and h are positive constants,

   (a) Find the positive value of t that will make $u(t) = 0$.

   (b) Find, $u'(t)$, when t is the value found in Part (a).

Chapter **5**

Derivatives

5.1 Domain, range, and derivatives

- Undefined and defined terms in mathematics.

- What is a function?

- What is the domain of a function?

- Slopes and Derivatives.

In mathematics, as well as in most other disciplines, we generally accept the notion that it is not possible to define every word without eventually circling back to the very word we were originally trying to define. Suppose you define the word "blue" as the color of a clear sky, then you may be asked to say what the color of a clear sky is and you would eventually be forced to say that it is blue. In order to break this problem of circular definitions, we are required to say that some words are commonly understood with no further attempt to define them.

5.2 Primitive terms

Certain words such as "number", "point", "set" are left as *undefined* or *primitive* terms. The readers are free to think of these abstract items as anything they want to just so long as their notions make sense in the definitions and problems using these terms.

This stage of the calculus, as we are learning it here, is called *two-dimensional* or "two variable" calculus and our notion of the word "point" is that it is an element in the xy-coordinate plane. Later in this book we will discuss *three-dimensional* calculus, where we take the word "point" to mean an element in xyz-coordinate space.

5.2.1 Definition: POINT SET

A *point set* is a set whose elements are points. Every point set contains at least one element. A point set may consist of exactly one single point or of any number of points. In two-dimensional calculus, a point set can contain points on either one of the coordinate axes or anywhere in the plane. A graph is a point set. Any table or equation that describes a relationship between two variables x and y can be used to describe a point set.

5.2.2 Definition: INTERVAL

If a and b are any two real numbers and $a < b$, then the *interval* $[a, b]$ is the set of all real numbers x such that $a \leq x \leq b$. The points a and b are called the end points of the interval $[a, b]$.

This definition says the interval $[a, b]$ is the set of all numbers between a and b *including the end points* a and b.

5.2.3 Definition: SEGMENT

If a and b are any two real numbers and $a < b$, then the *segment* (a, b) is the set of all real numbers x such that $a < x < b$. The points a and b are called the end points of the segment (a, b).

This definition says the segment (a, b) is the set of all numbers between a and b **not** *including the end points* a and b.

See Figure 5.1, depicting the interval $[a, b]$ (upper figure) and the segment (a, b).

Unfortunately, the notation for the point (a, b) in the plane is the same as the notation for the segment (a, b) on the number line. Usually the context in which we find this notation is sufficient to clarify which concept we are using.

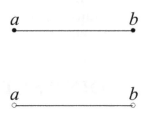

Figure 5.1 The interval $[a, b]$, and segment (a, b).

5.2.4 Questions: Intervals versus Segments

1. Is there some number in a segment that is greater than all other numbers in the segment?

2. Is there some number in an interval that is greater than all other numbers in the interval?

3. Is there a smallest number in a segment?

4. Is there a left-most point in an interval?

Answers:

1. By definition, the segment (a, b) includes all of the numbers, x, such that $a < x < b$. Suppose that there is a largest number in the segment. Call such a number t. Therefore, $a < t < b$. Now let r be the number halfway between t and b, then $r < b$, and since $t > a$, then so is $r > a$. Hence, r is a number in the segment (a, b), that is greater than the supposed largest number in (a, b). This contradiction proves that there is no largest number in the segment.

2. Yes, by definition b is a member of the interval $[a, b]$ and all other members are less than b.

3. No, because, if t is any member of (a, b) then $a < t$, so some other number r in the segment will be between t and a.

4. Yes, a is the left-most member of $[a, b]$.

5.3   What is a function?

5.3.1   Definition: FUNCTION

Suppose D is a number set, and that there is a number in D. The statement that, *f is a function whose domain is D* means that f is a collection of ordered pairs of numbers with the property that for each number x in D, there exists a number y such that the ordered pair (x, y) is in the collection f and no two pairs in f have the same first term.   The number set D is called the *domain* of the function.   If f is a function whose domain is D, then the *range* of f is the set of all numbers y such that (x, y) is in f.

If (x, y) is an ordered pair in f then, we write $y = f(x)$ and call this the *equation of the function f*. By the graph of a function we mean the set of all points whose coordinates satisfy the equation of the function. Thus, the graph of a function is a simple graph.

5.3.2   Example: The function $f(x) = \sqrt{9 - x^2}$ on [-3,3]

Let C be the set of all points $(x, f(x))$ such that $x \in [-3, 3]$ and $f(x) = \sqrt{9 - x^2}$, then C is the graph of the equation $y = f(x)$.   See Figure 5.2. Sometimes we just say C is the graph of $f(x)$.

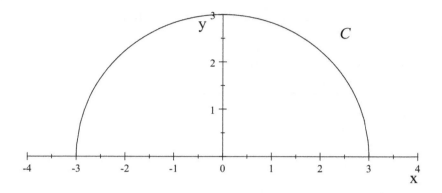

Figure 5.2 The simple graph C.

5.4   Domain restrictions

5.4.1   Natural Domain Restrictions

Sometimes the equation $y = f(x)$ of a function will necessarily *restrict* the domain of the function.

5.4.2   Problem: A restriction concerning square roots.

In Figure 5.2, the graph C of $f(x) = \sqrt{9 - x^2}$ shown on the interval $[-3, 3]$. Why can't the number $x = 4$ be in the domain?

Solution

Suppose that 4 could be in the domain of $f(x) = \sqrt{9 - x^2}$, then $f(4)$ would be $\sqrt{9 - 4^2} = \sqrt{9 - 16} = \sqrt{-7}$, not a real number. But numbers in the domain of a function must have real numbers that satisfy the equation of the function. This contradiction means that 4 is not in the domain.

Any restrictions on the domain caused by the equation itself, as we have just seen, is a *natural* restriction on the function. But sometimes a problem may require a restriction that comes from the application rather than the equation. We will call such restrictions *arbitrary* domain restrictions.

5.4.3   Arbitrary Domain Restrictions

In applied problems we may wish to impose certain conditions on the domain of a function, although the equation itself is not necessarily so restricted. For example if p is the population of New York City and w is the average wage of people there, then $f(p) = w \times p$ represents the total payroll. According to the equation p can be any number, so the domain is not restricted but we will want to use only positive numbers because it makes no sense to suppose p to be negative.

5.4.4   Questions about the domains of given functions

1. If $f(x)$ is the function $\sqrt{x}$, what values of x can be in its domain?

2. If $f(x) = \sqrt{25 - x^2}$, what values of x can be in its domain?

3. If $f(x) = \frac{1}{x}$, what values of x can be in its domain?

Answers:

A domain of $\sqrt{x}$, is any set that contains only non-negative real numbers; that is all $x \geq 0$.

A domain of $\sqrt{25 - x^2}$ is any set containing only the numbers x such that $-5 \leq x \leq 5$.

A domain of $1/x$ is any set of real numbers that do not include zero.

5.5 Slopes and derivatives

5.5.1 Definition: DERIVATIVE OF A FUNCTION

Let $f(x)$ be a function on a segment D, F be the graph of $y = f(x)$, and $a \in D$. Suppose that F has a tangent line with slope m at the point $(a, f(a))$, then m is called the *derivative* of f at a, and is denoted by $f'(a)$. In other words, *the derivative of a function at a point is the same as the slope of the graph at that point.*

5.5.2 Problem: What is the derivative at a point?

If D is a segment, $h(x)$ is a function on D and $a \in D$ such that the derivative of h exists at a, find $h'(a)$.

Solution

We find the derivative at a just as we find the slope at $(a, h(a))$,

$$h'(a) = \lim_{t \to a} \frac{h(t) - h(a)}{t - a}. \tag{5.1}$$

5.5.3 Problem: What is the derivative of x^4?

Let g be the function $g(x) = x^4$. If G is the graph of $y = g(x)$, find the derivative of g at a.

Solution

By definition:

$$
\begin{aligned}
g'(a) & = \lim_{p \to a} \frac{g(p) - g(a)}{p - a} \\
& = \lim_{p \to a} \frac{p^4 - a^4}{p - a}, \\
& = \lim_{p \to a} \frac{(p - a)(p^3 + p^2 a + pa^2 + a^3)}{p - a}, \\
& = \lim_{p \to a} p^3 + p^2 a + pa^2 + a^3, \\
g'(a) & = 4a^3.
\end{aligned}
$$

5.6   Graph of the sum of two functions

Suppose D is a domain and two functions $f_1(x)$ and $f_2(x)$ are defined on D, such that each has a derivative at each point of D. Let G_1 be the graph of the equation $y = f_1(x)$, let G_2 be the graph of the equation $y = f_2(x)$. We say that a graph G is the *sum of the two graphs G_1 and G_2*, when G is the graph of the equation $y = f_1(x) + f_2(x)$.

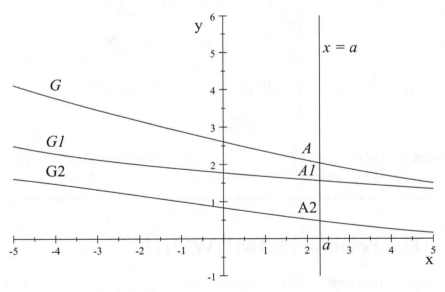

Figure 5.3 Graphs of $f_1(x)$, $f_2(x)$, and $f_1(x) + f_2(x)$.

The coordinates of points A_1, A_2, and A are

$$
\begin{aligned}
A_1 &= (a, f_1(a)). \\
A_2 &= (a, f_2(a)). \\
A &= (a, f(a)) \text{ or } (a, f_1(a) + f_2(a)).
\end{aligned}
$$

5.6.1 Problem: What is the derivative of $g(x) + h(x)$?

Let $g(x)$ and $h(x)$ be two functions defined on the same domain D, both having a derivative at each point of D. If $f(x)$ is the sum of $g(x)$ and $h(x)$, then show that $f'(x)$ is the sum of $g'(x)$ and $h'(x)$.

Solution

We are given that $f(x) = g(x) + h(x)$, so by definition of derivative:

$$
\begin{aligned}
f'(x) &= \lim_{t \to x} \frac{f(x) - f(t)}{x - t} \\
f'(x) &= \lim_{t \to x} \frac{g(x) + h(x) - (g(t) + h(t))}{x - t}, \\
f'(x) &= \lim_{t \to x} \frac{g(x) - g(t) + h(x) - h(t)}{x - t}, \\
f'(x) &= \lim_{t \to x} \left(\frac{g(x) - g(t)}{x - t} + \frac{h(x) - h(t)}{x - t} \right), \\
f'(x) &= \lim_{t \to x} \frac{g(x) - g(t)}{x - t} + \lim_{t \to x} \frac{h(x) - h(t)}{x - t}, \\
f'(x) &= g'(x) + h'(x).
\end{aligned}
$$

Therefore $(g(x) + h(x))' = g'(x) + h'(x)$. The derivative of the sum of two functions is the sum of their derivatives.

5.7 Exercise 5 HOMEWORK

1. Given a function f with equation $f(x) = \frac{-1}{(x-2)(x-3)}$ and whose domain D is the segment $(2, 3)$, draw the graph of f.

2. Write the expression $(p - a)$ as the difference of two cubes.

3. Let $f(x) = x^{1/3}$

   (a) What is the natural domain of f?

   (b) If P is point of the graph of $f(x)$, with abscissa p, what is the ordinate of P?

   (c) If A is a point of the graph of $f(x)$, with abscissa a, find the slope of the secant line PA.

   (d) In your answer to (c) write the denominator $(p - a)$ in such a way that you can cancel certain terms after which you can let $p \to a$ without making the denominator $\to 0$.

4. Let n be any positive integer, and let G be the graph of $f(x) = x^n$. Find the derivative of $f(x)$.

5. If n is any positive integer find the derivative of $g(x) = x^{1/n}$.

6. Let $g(x) = x^{-2}$

   (a) If P and A are two points of the graph G of $g(x)$, with abscissas p and a respectively, find the slope of the line AP.

   (b) Simplify your answer in (a) so that you can let $p \to a$ to find the slope of G at the point A.

7. Let $h(x) = x^5 + x^4 + x^3 + x^2 + x^1 + 1 + x^{-1} + x^{-2}$. Find $h'(x)$.

8. For the following functions $f(x)$ and $g(x)$, solve the following problems.

   (a) If $f(x) = x^{100}$, find $f'(x)$; if $g(x) = 5f(x)$, find $g'(x)$. If $h(x) = Cf(x)$ for any constant C, find $h'(x)$.

   (b) Prove that the derivative of the difference of two functions is the difference in their derivatives.

9. If $f(x) = x^3$, and $g(x) = x^4$ and $h(x) = f(x)g(x)$, is it true that $h'(x) = f'(x)g'(x)$?

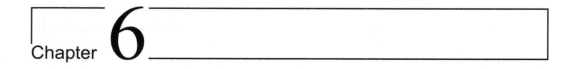

Chapter 6

Derivative Formulas

6.1 Derivatives of combinations of functions

- What are the derivative formulas for power functions?

- What are the derivative formulas for sums and differences of functions?

- What are the derivative formulas for products and quotients of functions?

6.1.1 Definition: POWER FUNCTION

Let D be a subset of the positive numbers and a be any real number. If $x \in D$, and x^a exists, then the collection of ordered pairs: $\{(x, x^a)\}$ is a *power function* on D. Its equation is $y = x^a$. There are infinitely many such functions, x^2, $x^{3/4}$, $x^{3.14}$, x^{-7}. So, anything of the form x^a is called a power function.

6.1.2 Examples: Graphs of Power Functions

If $a = 1$, the power function is x^1, the equation is $y = f(x) = x$ and the graph is a straight line with constant slope, $f'(x) = 1$ for all x; see Figure 6.1.

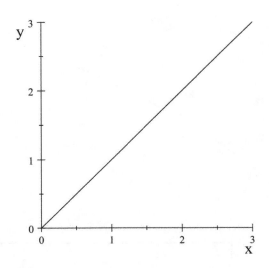

Figure 6.1 Graph of the power function x^1.

If $a > 1$, the equation is $y = f(x) = x^a$ and has a graph whose slope is positive and increasing.   See Figure 6.2.

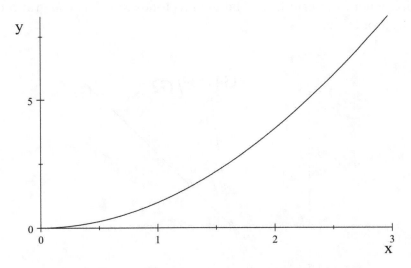

Figure 6.2 Graph of $f(x) = x^a$ for $a > 1$.

If a is positive but less than 1, the graph of the power function x^a is increasing.   It has a positive slope, but the slope itself is deceasing, as in Figure 6.3.

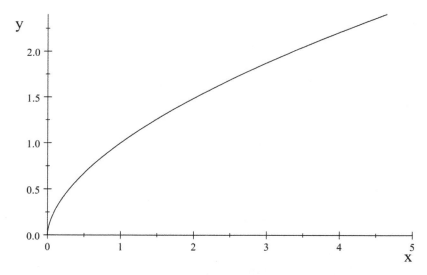

Figure 6.3 Graph of $f(x) = x^a$ for $0 < a < 1$.

Some power functions are: $y = x^1$, $y = x^2$, $y = x^3$, $y = x^4$, etc. See Figure 6.4, where the graphs are labeled as follows: $G1$ is the graph of x^1, $G2$ is the graph of x^2, etc.

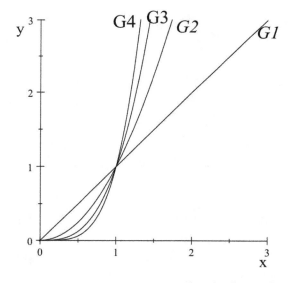

Figure 6.4 Power Functions, x^1, x^2, x^3, and x^4.

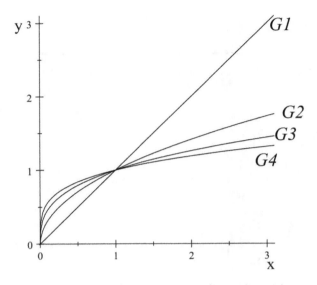

Figure 6.5 Power functions x, $x^{1/2}$, $x^{1/3}$, $x^{1/4}$.

In Figure 6.5, we see examples of the power functions whose equations are $y = x^{1/n}$. Here, $G1$ is the graph of $y = x$, $G2$ is the graph of $y = x^{1/2}$, $G3$ is the graph of $y = x^{1/3}$ and $G4$ is the graph of $y = x^{1/4}$.

6.2   Formulas for derivatives of power functions

In the Problems 4 and 5 in the homework Exercise 5, we found the following derivatives:

$$\text{If } f(x) = x^n, \text{ then } f'(x) = nx^{n-1}.$$
$$\text{If } f(x) = x^{\frac{1}{n}}, \text{ then } f'(x) = \frac{1}{n}x^{\frac{1}{n}-1}.$$

Now let us look at power functions, x^a in which the power a is a negative integer.

6.2.1   Problem: Graphs of $y = x^{-n}$

Sketch the graphs of the functions $y = x^{-1}$, $y = x^{-2}$, $y = x^{-3}$, $y = x^{-4}$, on the segment $(0, 3)$.

Solution

Let $G1$ be the graph of $y = x^{-1}$, $G2$ be the graph of $y = x^{-2}$, etc. as in Figure 6.6

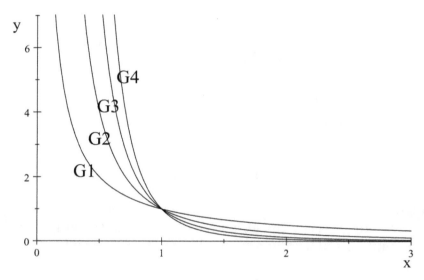

Figure 6.6 Graphs of $y = x^{-1}$, $y = x^{-2}$, $y = x^{-3}$, $y = x^{-4}$.

Next, we ask the question will the derivative of x^{-n} look like?

6.2.2 Problem: Derivative of x^{-n}

Let $f(x) = x^{-n}$, where n is any positive integer and $x \neq 0$. Use the definition of derivative to find $f'(x)$. Simplify your answer so that it will resemble the derivative of x^n.

Solution

Let $f(x) = x^{-n}$, then by definition of derivative

$$\begin{aligned} f'(x) &= \lim_{t \to x} \frac{f(t) - f(x)}{t - x}, \\ &= \lim_{t \to x} \frac{t^{-n} - x^{-n}}{t - x}. \end{aligned} \qquad (6.1)$$

In order to simplify the computation, let us work with the *numerator* of Equation (6.1).

$$\begin{aligned} t^{-n} - x^{-n} &= \frac{1}{t^n} - \frac{1}{x^n}, \\ &= \frac{x^n - t^n}{t^n x^n}. \end{aligned}$$

Now Equation (6.1) becomes

$$\begin{aligned} f'(x) &= \lim_{t \to x} \frac{x^n - t^n}{(t - x)t^n x^n} \\ &= \lim_{t \to x} -\frac{t^n - x^n}{(t - x)t^n x^n}, \\ &= \lim_{t \to x} -\frac{t^n - x^n}{t - x} \frac{1}{t^n x^n}, \\ &= \lim_{t \to x}(-1)\frac{t^n - x^n}{t - x} \lim_{t \to x} \frac{1}{t^n x^n}. \end{aligned} \qquad (6.2)$$

Now, look at the limits in (6.2). We already know that

$$\lim_{t \to x}(-1)\frac{t^n - x^n}{t - x} = (-1)nx^{n-1}.$$

What about

$$\lim_{t \to x} \frac{1}{t^n x^n}?$$

It is $1/x^{2n}$. So the limit in (6.2) becomes $-nx^{n-1}/x^{2n}$, which simplifies as follows

$$-n\frac{x^{n-1}}{x^{2n}} = -nx^{-n-1}.$$

Therefore, when $f(x) = x^{-n}$, $f'(x) = -nx^{-n-1}$.

Now what about $x^{m/n}$? In the homework, Problem 1 Exercise 6, you will get a chance to find $y'(x)$ when $y(x) = x^{m/n}$.

6.2.3 Test your intuition: Make a guess right now.

If $y(x) = x^{m/n}$, what is your guess about $y'(x)$?

Answer

$$y'(x) = \frac{m}{n}x^{\frac{m}{n}-1}$$

6.2.4 Question: What is the derivative of x^a? Make a guess right now.

Based upon the preceding examples, what is the derivative of the power function x^a, in general, where a is *any* constant, not necessarily a rational number?

Answer

If $f(x) = x^a$, then our previous calculations seem to indicate that the general formula for the derivative would be $f'(x) = ax^{a-1}$. This is true, however we did not prove it (in general) here. The proof will be revealed later when we find the derivatives of the logarithms.

6.3 Formula for the derivative of the product

6.3.1 Problem: What is the derivative of $xg(x)$?

Let $f(x) = x$, on the domain D, and $g(x)$ be any other function that has a derivative on D. Let $h(x) = xg(x)$. Use the definition of derivative to find $h'(x)$.

Attempted Solution

By the definition of derivative

$$\begin{aligned} h'(x) &= \lim_{t \to x} \frac{h(t) - h(x)}{t - x}, \\ h'(x) &= \lim_{t \to x} \frac{tg(t) - xg(x)}{t - x}. \end{aligned} \qquad (6.3)$$

We're stuck.   We can't let $t \to x$ yet because we don't have a way to get rid of the $t - x$ in the denominator.   This is a good time to recite the following famous poem:

6.3.2   The Old Mathematician's Trick

> *THE OLD MATHEMATICIAN'S TRICK*
> *Look at what you've got.*
> *Look at what you'd rather have.*
> *Change what you've got*
> *To what you'd rather have,*
> *And make the proper adjustment.*

What we've got is $tg(t) - xg(x)$.   We would rather have an expression we can factor; something that would eventually produce the factors $t - x$, and $g(t) - g(x)$ in the numerator.

6.3.3   Problem: Changing what we've got

Find a way to change $tg(t) - xg(x)$ so as to get $tg(t) - tg(x)$, among some other things.

Solution

Let us insert (by subtracting) the term $tg(x)$ and then adjust for this by adding it.   So

$$tg(t) - xg(x) = tg(t) - tg(x) + tg(x) - xg(x). \qquad (6.4)$$

6.3.4   Problem: How do we finish this?

Use Equations (6.3) and (6.4) to finish the problem of finding the derivative of $h(x) = xg(x)$ by the definition of derivative.

Solution

Starting with Equation (6.3), and using Equation (6.4), we have:

$$
\begin{aligned}
h'(x) &= \lim_{t \to x} \frac{tg(t) - xg(x)}{t - x} \\
&= \lim_{t \to x} \frac{tg(t) - tg(x) + tg(x) - xg(x)}{t - x}, \\
&= \lim_{t \to x} \frac{tg(t) - tg(x)}{t - x} + \frac{tg(x) - xg(x)}{t - x}, \\
&= \lim_{t \to x} \frac{t}{1} \frac{g(t) - g(x)}{t - x} + \lim_{t \to x} \frac{t - x}{t - x} \frac{g(x)}{1}, \\
&= \lim_{t \to x} \frac{t}{1} \lim_{t \to x} \frac{g(t) - g(x)}{t - x} + \lim_{t \to x} \frac{1}{1} \lim_{t \to x} \frac{g(x)}{1}, \\
h'(x) &= xg'(x) + g(x).
\end{aligned}
$$

Therefore

$$(xg(x))' = xg'(x) + g(x). \qquad (6.5)$$

To find the derivative of the product of x and $g(x)$, multiply x by the derivative of g, then add the derivative of x times $g(x)$. We will leave it as a homework problem (Problem 2, Exercise 6) for you to find that derivative of the product $f(x)g(x)$ of two functions. In the meantime, let us suppose the homework problem has been solved; you will note that this formula may be written in words as follows

> The derivative of the product of two functions is the first times the derivative of the second plus the second times the derivative of the first.

The advantage in saying it this way is that it takes about as long to say it as it does to write it; so when you are writing out the derivative and saying it at the same time, you come out even. Also, it is better to learn this verbal form rather than a form that uses notation such as f and g because the functions you are working on may not be called f and g.

6.3.5   Example: What is the derivative of $x^3 z(x)$

Given that $z(x)$ is a function that has a derivative, verbally state the process of finding the derivative of the product $x^3 z(x)$ while writing it.

Solution

Say "The first times the derivative of the second" while writing $x^3 z'(x)$. Then say "Plus" while writing $+$. Next say "The second times the derivative of the first", while writing $z(x)\ 3x^2$.

Thus,

$$(x^3\ z(x))' = x^3\ z'(x) + z(x)\ 3x^2.$$

6.3.6   Example: What is the derivative of $x^5(x^2 + 4x^{-3})$?

Find the derivative of $x^5(x^2 + 4x^{-3})$

Solution

"The first times the derivative of the second", $x^5(2x - 12x^{-4})$. "Plus", $+$,

"The second times the derivative of the first": $(x^2 + 4x^{-3})5x^4$.

That is,

$$(x^5(x^2 + 4x^{-3}))' = x^5(2x - 12x^{-4}) + (x^2 + 4x^{-3})5x^4.$$

Once you know how to find the derivative of the power functions, and how to find the derivative of the sum and difference of two functions, you can find the derivative of any sum or difference of power functions. Thus, for example, you can find the derivative of any polynomial. And knowing the formula for finding the derivative of the product of functions opens up many more possibilities.

6.3.7 Example: Derivative of $(2x^{100} - 3x + 10)(x^3 + x^2 + x + 1)$

Find the derivative of $(2x^{100} - 3x + 10)(x^3 + x^2 + x + 1)$

Solution

It is: $(2x^{100} - 3x + 10)(3x^2 + 2x + 1) + (x^3 + x^2 + x + 1)(200x^{99} - 3)$

6.4 Exercise 6 HOMEWORK

1. If m and n are positive integers and D is a domain on which the power function $x^{\frac{m}{n}}$ exists, use the definition of derivative to show that $(x^{\frac{m}{n}})' = \frac{m}{n} x^{\frac{m}{n}-1}$. You may use the fact that in Problem 5 of Exercise 5, you proved that $(x^{\frac{1}{n}})' = \frac{1}{n} x^{\frac{1}{n}-1}$.

2. Let two functions $f(x)$ and $g(x)$ be defined on the same domain D and each having a derivative on D. Prove, from the definition of derivative, that
$$(f(x)g(x))' = f(x)g'(x) + g(x)f'(x).$$

3. Given two functions $f(x)$ and $g(x)$ defined on the same domain D and such that their respective derivatives, $f'(x)$ and $g'(x)$ exist on D, and assume $g(x) \neq 0$ on D, use the definition of derivative to find $\left(\frac{f(x)}{g(x)}\right)'$.

4. Complete the following statement in order to verbally express the formula for the derivative of the quotient of two functions. "The derivative of the quotient of two functions is..".

5. Use the formula for the quotient of two functions to show that $\left(\frac{1}{f(x)}\right)' = \frac{-f'(x)}{f(x)^2}$.

6. For the function $g(x) = \frac{x-1}{(x-2)(x-3)}$,

   (a) Sketch the graph of $y = g(x)$.

(b) Find the derivative $g'(x)$.

7. For the function $h(x) = \frac{1}{x}$:

    (a) Sketch the graph of $y = h(x)$ on $(0, 5]$.

    (b) Sketch the graph of $y = h'(x)$ on $(0, 5]$.

7

The Limit of sin(x)/x as x → 0

7.1 Radians and trigonometric functions

7.1.1 Questions we want to investigate

- What is the indeterminate form 0/0?

- What happens to $\sin(x)/x$ as $x \to 0$?

- Measuring angles in real numbers.

- Trigonometric functions defined on a right triangle.

- Trigonometric functions on the unit circle.

- What is the derivative of $\sin(x)$?

7.2 What is the indeterminate form $0/0$?

When we use the definition to find the derivative of a function, $f(x)$ at some point a,we are finding out what happens to the following quotient as $x \to a$,

$$\frac{f(x) - f(a)}{x - a}.$$

If we try to immediately substitute $x = a$ into this expression, we will arrive at $\frac{0}{0}$ which is called an "indeterminate form," because its value cannot be

determined. It is a symbolic monstrosity that is meaningless unless some
thing can be done to "simplify" the expression before letting x become a.
Up until now, we were able to carry out this simplification by re-writing the
numerator so that we could factor the numerator or denominator in a way
that let us cancel the $(x - a)$ in the denominator, then let $x \to a$, *without
getting* 0/0.

7.2.1 Example: What happens to $\frac{\sqrt[3]{x}-\sqrt[3]{a}}{x-a}$ as $x \to a$?

Find out what happens to $\frac{\sqrt[3]{x}-\sqrt[3]{a}}{x-a}$ as $x \to a$. If we try to substitute in $x = a$,
we get $\frac{0}{0}$. What neat trick did we do before to fix this situation? One idea
was to re-write $x - a$ as

$$x - a = (\sqrt[3]{x})^3 - (\sqrt[3]{a})^3.$$

We can then factor this as the difference of two cubes. Here, for example,
we get

$$(\sqrt[3]{x})^3 - (\sqrt[3]{a})^3 = (\sqrt[3]{x} - \sqrt[3]{a})\left((\sqrt[3]{x})^2 + \sqrt[3]{x}\sqrt[3]{a} + (\sqrt[3]{a})^2\right).$$

So, the fraction becomes

$$\frac{\sqrt[3]{x} - \sqrt[3]{a}}{x - a} = \frac{(\sqrt[3]{x} - \sqrt[3]{a})}{(\sqrt[3]{x} - \sqrt[3]{a})((\sqrt[3]{x})^2 + \sqrt[3]{x}\sqrt[3]{a} + (\sqrt[3]{a})^2)}.$$

If we cancel the factor $(\sqrt[3]{x} - \sqrt[3]{a})$ from both the numerator and denominator, we will get

$$\frac{\sqrt[3]{x} - \sqrt[3]{a}}{x - a} = \frac{1}{((\sqrt[3]{x})^2 + \sqrt[3]{x}\sqrt[3]{a} + (\sqrt[3]{a})^2)}.$$

To find the limit as $x \to a$, we simply let $x = a$, without getting $\frac{0}{0}$.
Instead, we get

$$\frac{1}{3(\sqrt[3]{a})^2}.$$

In other words, the derivative of $x^{1/3}$ at $x = a$ is $\frac{1}{3}a^{-2/3}$.

The indeterminate form, $\frac{0}{0}$, is one of the single most difficult conundrums
encountered in beginning calculus. An indeterminate form is sometimes
called an indeterminacy, and every formula that we obtain from the definition of derivative, will require us to find some clever way to get rid of the
indeterminacy.

7.3 What happens to $\sin(x)/x$ as $x \to 0$?

We are about to meet one of the most problematic indeterminate forms that we have encountered yet. It occurs when we try to find the derivative of $f(x) = \sin(x)$ from the definition of derivative. What we will eventually need to do is to find out what happens to $\sin(x)/x$, as $x \to 0$. It clear that we can't just substitute in $x = 0$. Do you see why not?

In order to solve this, it will be necessary for us to review a little trigonometry. We start with measuring angles.

7.4 Measuring arcs and angles

7.4.1 Definition: CIRCULAR ARC

Let A and B be two points, not diametrically opposite of each other, on the circumference of a circle, the minor circular arc $\overset{\frown}{AB}$ is the smaller part of the circumference between A and B. The major arc is the larger fraction of the circumference. See Figure 7.1.

7.4.2 Definition: CENTRAL ANGLE

Let $\overset{\frown}{AB}$ be the minor arc of a circle whose center is O, and let ψ be the angle $\angle AOB$, then ψ is the central angle. See Figure 7.1.

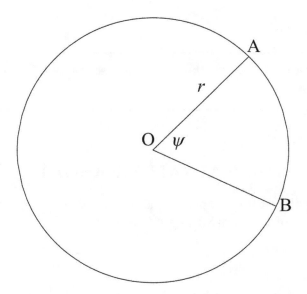

Figure 7.1 The central angle $\angle AOB$.

7.4.3   Definition: MEASURE OF A CENTRAL AN-GLE

Given any circular arc $\widehat{AB}$, we can assign a measure to the central angle. When the arc is the whole circle, the central angle can be said to have a measure of 360°. Or, if we take our measurements in another unit, called "radians", the central angle for a complete circle is 2π. In other words,

$$2\pi \text{ radians} = 360°. \tag{7.1}$$

When the arc length is **not** the whole circle such as the arc $\widehat{AB}$ in Figure 7.1, then the measurement, ψ, of the central angle will be related to some fraction of the whole circle. That is, some fraction of 360°, or some fraction of 2π.   Thus, for example, when the arc length is one-fourth the whole circumference, then the central angle has a measure of $\frac{1}{4}2\pi = \frac{\pi}{2}$ or in degrees, $\frac{1}{4}360° = 90°$.

7.4.4   Question: How many degrees in a radian?

Approximately how many degrees are in one radian?

Answer

From Equation (7.1), we have 2π radians $= 360°$, therefore 1 radian $= 180°/\pi \approx 57.3°$

7.4.5   Definition: RADIAN MEASURE

If $\mathcal{C}$ is a circle with radius r and $\overset{\frown}{AB}$ is an arc with length s on $\mathcal{C}$, then the radian measure x of the central angle is s/r.

$$x \;=\; \frac{s}{r}, \text{ or}$$

$$\text{Central angle (in radians)} \;=\; \frac{\text{arc length}}{\text{radius}}.$$

Thus, if the length, s, of the arc $\overset{\frown}{AB}$ is equal to the radius, r, then the central angle $\angle AOB$ is r/r or 1 radian.

7.4.6   Problem: If $r = 2.47$ and $x = 1.33$ find s

If a circle with radius 2.47 inches has a central angle $\angle AOB$ whose radian measure is 1.33, find the length of the arc $\overset{\frown}{AB}$.

Solution

Here the central angle measured in radians is $x = 1.33$, and the radius of the circle $r = 2.47$. Let s be the arc length, then $x = s/r$ or

$$\begin{aligned}
\text{(Arc length)} \;&=\; \text{(Central angle)} \times \text{(Radius)}, \\
s \;&=\; x \times r, \\
s(\text{inches}) \;&=\; 1.33 \times 2.47(\text{inches}), \\
s \;&=\; 3.2651 \text{ inches.}
\end{aligned}$$

7.5   Functions defined on a right triangle

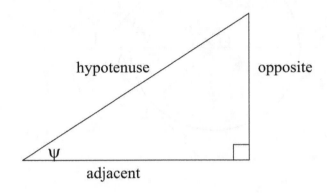

Figure 7.2 Trig Functions on a right triangle.

$$\sin(\psi) = \tfrac{opp}{hyp}.$$
$$\cos(\psi) = \tfrac{adj}{hyp}.$$
$$\tan(\psi) = \tfrac{opp}{adj}.$$
$$\tan(\psi) = \tfrac{\sin(\psi)}{\cos(\psi)}.$$
$$\sec(\psi) = \tfrac{1}{\cos(\psi)}.$$
$$\csc(\psi) = \tfrac{1}{\sin(\psi)}.$$
$$\cot(\psi) = \tfrac{1}{\tan(\psi)}.$$

7.6   Functions defined on the unit circle

Consider a circle with radius $r = 1$.

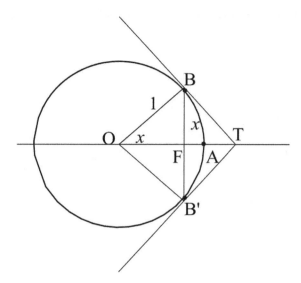

Figure 7.3 Unit circle with $\angle BOA = x$.

In this figure O is the center, the length of the arc $\overset{\frown}{BA}$ is x, the radius $r = 1$, the points A, B, and B' are on the circle. Thus $\overline{OA} = \overline{OB} = \overline{OB'} = 1$. The line TB is tangent to the circle at B, and the line TB' is tangent to the circle at B'.

7.6.1 Question: In a unit circle, why does the arc length = the angle measure?

In Figure 7.3, why is the length of the arc $\overset{\frown}{BA}$ the same as the measure of the angle $\angle BOA$?

Answer

Because the radius is $r = 1$; the arc length, s is $\overset{\frown}{BA}$; and the radian measure, x of the central angle is

$$x = \frac{s}{r} \text{ or } \frac{\overset{\frown}{BA}}{1}.$$

The following trigonometric functions are defined in terms of the triangles in Figure 7.3

$$\begin{aligned} \sin(x) &= \overline{FB} \\ \tan(x) &= \overline{BT} \end{aligned}$$

7.6.2   Question: In Figure 7.3 why is $\overline{BT} = \tan(x)$?

Why is the segment $\overline{BT} = \tan(x)$?

Answer

The triangle $\triangle OBT$ is a right triangle because $\overline{BT} \perp \overline{OB}$, and since $\overline{OB}$ is the side adjacent to x and $\overline{BT}$ is the side opposite to x, then $\tan(x) = \frac{\overline{BT}}{\overline{OB}}$.

We make the following observations, for $x < \pi/2$.

1. The segment $\overline{BB'} = \overline{BF} + \overline{FB'} = 2\sin(x)$.

2. The arc $\overset{\frown}{BB'}$ is $2x$.

3. The line segment $\overline{BB'}$ is shorter than the arc $\overset{\frown}{BB'}$. So $2\sin(x) < 2x$.

4. The arc $\overset{\frown}{BB'}$ is shorter than the distance from B to T plus the distance from T to B', or

$$\begin{aligned} \overset{\frown}{BB'} &< BT + TB', \\ 2x &< 2\tan(x). \end{aligned}$$

Thus, for all x such that $x > 0$ and $x < \pi/2$, we have the inequality

$$2\sin(x) \quad < \quad 2x < 2\tan(x), \tag{7.2}$$

$$\sin(x) \quad < \quad x < \frac{\sin(x)}{\cos(x)}. \tag{7.3}$$

Divide by $\sin(x)$ which is positive for the values of x being considered here, and we get

$$1 < \frac{x}{\sin(x)} < \frac{1}{\cos(x)}. \tag{7.4}$$

This means that $x/\sin(x)$ is always between 1 and $1/\cos(x)$.

7.6.3   Questions: What happens as $x \to 0$?

1. What happens to $\cos(x)$ as $x \to 0$?
2. What happens to $1/\cos(x)$ as $x \to 0$?
3. What happens to 1 as $x \to 0$?
4. What happens to $\frac{\sin(x)}{x}$ as $x \to 0$?

Answers

---

1. $\cos(x) \to 1$ as $x \to 0$.
2. $1/\cos(x) \to 1$ as $x \to 0$.
3. 1 stays 1 as $x \to 0$.
4. By Inequality (7.4), $x/\sin(x)$ is always closer to 1 than $1/\cos(x)$.
So, as $1/\cos(x) \to 1$, so must $x/\sin(x) \to 1$. From this we conclude

$$\lim_{x \to 0} \frac{\sin(x)}{x} = 1. \tag{7.5}$$

7.6.4   Problem: Find $\lim_{x \to 0} \frac{\sin(3x)}{x}$.

Find

$$\lim_{x \to 0} \frac{\sin(3x)}{x}. \tag{7.6}$$

Solution

First, let us multiply numerator and denominator by 3 and write

$$\frac{\sin(3x)}{x} = \frac{3\sin(3x)}{3x}.$$

Now let $3x = t$; hence we have

$$\frac{3\sin(t)}{t}.$$

As $x \to 0$, $3x \to 0$; that is $t \to 0$

$$\lim_{x \to 0} \frac{\sin(3x)}{x} = \lim_{x \to 0} \frac{3\sin(3x)}{3x} = \lim_{t \to 0} \frac{3\sin(t)}{t} = 3.$$

7.6.5   Problem: Find $\lim_{t \to a} \frac{\sin(t-a)}{t-a}$.

Find

$$\lim_{t \to a} \frac{\sin(t - a)}{t - a}.$$

Solution

Let $x = t - a$, then as $t \to a$, $x \to 0$. So, the problem becomes find

$$\lim_{x \to 0} \frac{\sin(x)}{x}.$$

The answer is 1.

7.7   The derivative of sin(x)

We are now ready to try to find the derivative of $\sin(x)$.

7.7.1   Problem: Find the derivative of $\sin(x)$ at $x = a$

If a is any real number, and $f(x) = \sin(x)$, use the definition of the derivative to find $f'(a)$.

Solution

By definition

$$
\begin{aligned}
f'(a) &= \lim_{t \to a} \frac{f(t) - f(a)}{t - a} \\
&= \lim_{t \to a} \frac{\sin(t) - \sin(a)}{t - a}.
\end{aligned}
\tag{7.7}
$$

But we don't have a way to get rid of the indeterminacy $\frac{0}{0}$ unless we can find some useful trig identity, preferably one that makes use of $\frac{\sin(x)}{x}$, since we already know how to deal with that. We temporarily abandon this attempt in favor of searching for an appropriate trig identity.

*** Solution abandoned ***

Let's resort to the old mathematicians trick. Look at what we've got: $\frac{\sin(t)-\sin(a)}{t-a}$; what we would rather have is something like $\frac{\sin(t-a)}{t-a}$. What kind of adjustment can we make? Let us try changing t to a sum of two terms such as $p+q$ and a to a difference of two terms such as $p-q$. Then we can write $\sin(t)$ as $\sin(p+q)$, and use the identity: $\sin(p+q) = \sin(p)\cos(q) + \cos(p)\sin(q)$. And $\sin(a)$ becomes $\sin(p-q) = \sin(p)\cos(q) - \cos(p)\sin(q)$.

Solution re-started

Let $t = p+q$ and $a = p-q$, then

$$
\begin{aligned}
\sin(t) - \sin(a) &= \sin(p+q) - \sin(p-q) \\
&= \sin(p)\cos(q) + \cos(p)\sin(q) \\
&\quad - \sin(p)\cos(q) + \cos(p)\sin(q), \\
&= 2\cos(p)\sin(q).
\end{aligned}
$$

Now $t - a = 2q$, and since $t \to a$, then $t - a \to 0$. Also notice that $t - a \to 0$ implies that $q \to 0$. In addition, $p = a + q$, therefore $p \to a$, as $q \to 0$

This means that we can rewrite (7.7) as follows

$$
\begin{aligned}
\lim_{t \to a} \frac{\sin(t) - \sin(a)}{t - a} &= \lim_{2q \to 0} \frac{2\cos(p)\sin(q)}{2q} \\
&= \lim_{q \to 0} \frac{\cos(p)\sin(q)}{q}, \\
&= \lim_{q \to 0} \frac{\cos(p)}{1} \times \lim_{q \to 0} \frac{\sin(q)}{q}, \\
&= \cos(a) \times 1.
\end{aligned}
$$

Therefore,

$$
f'(a) = \lim_{t \to a} \frac{\sin(t) - \sin(a)}{t - a} = \cos(a). \tag{7.8}
$$

Thus, we have shown that the derivative of $\sin(x)$ at any point a is $\cos(a)$. Using x, in general as the value for a, we are able to write this as the following important formula:

$$(\sin(x))' = \cos(x).$$

From this one formula, we can readily find the derivatives of other trig functions.

7.8   Exercise 7 HOMEWORK

1. Find
$$\lim_{x \to 0} \frac{\sin(x)}{\sqrt{x}}.$$

2. Find
$$\lim_{x \to 0} \frac{\sin(\sqrt{x})}{x}.$$

3. If $f(x) = \cos(x)$, use the definition of derivative to find $f'(x)$.

4. If $f(x) = \tan(x)$, use a trig identity to find $f'(x)$.

5. Use a trig identity to find the derivative of $\csc(x)$.

6. Use a trig identity to find the derivative of $\sec(x)$.

7. Use a trig identity to find the derivative of $\cot(x)$.

8. Find the derivative of $\sin(x^3)$.

9. Find the derivative of $\sin(\sqrt{x})$.

Chapter **8**

The Chain Rule

8.1 Questions about derivatives.

Here are some questions we want to investigate.

- What is the derivative of $(f(x))^n$?

- How do you find the derivative of $f(\sin(x))$ and of $\sin(f(x))$?

- What is derivative notation?

- How do you find the derivative of composite functions $f(g(x))$?

8.2 Derivative of $(f(x))^n$

8.2.1 Problem: For any integer n find $y'(x)$, when $y(x) = (f(x))^n$

If n is any integer and $f(x)$ is any function that has a derivative at some point $(a, f(a))$, and if $y(x) = (f(x))^n$ on some segment containing a, prove that $y'(a) = n(f(a))^{n-1}f'(a)$.

Solution

$$y(x) = (f(x))^n$$

$$y'(a) = \lim_{x \to a} \frac{(f(x))^n - (f(a))^n}{x - a},$$

$$= \lim_{x \to a} \frac{(f(x))^n - (f(a))^n}{f(x) - f(a)} \times \frac{f(x) - f(a)}{x - a},$$

$$= \lim_{x \to a} \frac{(f(x))^n - (f(a))^n}{f(x) - f(a)} \times \lim_{x \to a} \frac{f(x) - f(a)}{x - a},$$

$$= \lim_{f(x) \to f(a)} \frac{(f(x))^n - (f(a))^n}{f(x) - f(a)} \times \lim_{x \to a} \frac{f(x) - f(a)}{x - a},$$

$$= n(f(x))^{n-1} \times f'(a).$$

The above formula illustrates how we can find the derivative of a "function of a function". In this, case $f(x)$ which is a function of x was raised to the nth power. In other words, $f(x)$ is "the inside function " and the nth power is "the outside function". We can state the formula verbally as follows: "The derivative of the nth power of $f(x)$ with respect to x is the derivative of the nth power of $f(x)$ with respect to $f(x)$ times the derivative of $f(x)$ with respect to x". In general, this says "The derivative of a function of a function is the derivative of the *outside function* with respect to the *inside function* times the derivative of the *inside* function with respect to its variable". This is a special case of the *chain rule* formula.

Now let's derive some particular chain rule formulas involving trig functions.

8.3   Derivative of $f(\sin(x))$ and $\sin(f(x))$

8.3.1   Example: Derivative of $\sin^2(x)$

Suppose we needed the derivative of $(\sin(x))^2$, how could we find it?

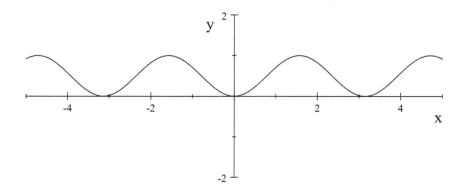

Figure 8.1 Graph of $y(x) = (\sin(x))^2$.

One way is to recognize that $(\sin(x))^2$ is the product of $\sin(x)$ times $\sin(x)$. So, we could use the formula for the derivative of a product

$$
\begin{aligned}
(\sin(x) \times \sin(x))' &= \sin(x) \times (\sin(x))' + \sin(x) \times (\sin(x))' \\
&= 2\sin(x)\cos(x).
\end{aligned}
$$

But here is another approach, one that provides more insight to the problem of finding the derivative of a function of a function such as $f(\sin(x))$. If $y(x) = (\sin(x))^2$ then

$$
\begin{aligned}
y'(x) &= \lim_{t \to x} \frac{(\sin(t))^2 - (\sin(x))^2}{t - x} \\
&= \lim_{t \to x} \frac{(\sin(t))^2 - (\sin(x))^2}{\sin(t) - \sin(x)} \times \frac{\sin(t) - \sin(x)}{t - x}.
\end{aligned}
\tag{8.1}
$$

We got this from multiplying numerator and denominator by $\sin(t) - \sin(x)$.

Now as $t \to x$, we see that $\sin(t) \to \sin(x)$, so we can re-write the limits in (8.1) as follows

$$
y'(x) = \lim_{\sin(t) \to \sin(x)} \frac{(\sin(t))^2 - (\sin(x))^2}{\sin(t) - \sin(x)} \times \lim_{t \to x} \frac{\sin(t) - \sin(x)}{t - x}.
\tag{8.2}
$$

In (8.2), let us temporarily call $\sin(t)$ and $\sin(x)$ in the first fraction p and q, respectively; but leave the second fraction as is.   Then as $\sin(t) \to \sin(x)$, $p \to q$.   Thus

$$y'(x) = \lim_{p \to q} \frac{p^2 - q^2}{p - q} \times \lim_{t \to x} \frac{\sin(t) - \sin(x)}{t - x}. \tag{8.3}$$

In (8.3), the first fraction has $2q$ for its limit because it is the definition of the derivative of x^2 at q.   The second fraction has $\cos(x)$ for its limit because it is the definition of the derivative of $\sin(x)$ at x.   Therefore

$$y'(x) = 2q \cos(x) = 2 \sin(x) \cos(x).$$

See Figure (8.2) for the derivative of $(\sin(x))^2$.

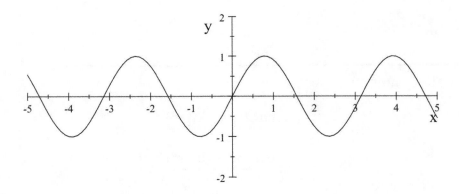

Figure 8.2 Graph of $2 \sin(x) \cos(x)$, the derivative of $(\sin(x))^2$.

Now let us look at $\sin(x^2)$ and find its derivative.   See Figure (8.3).

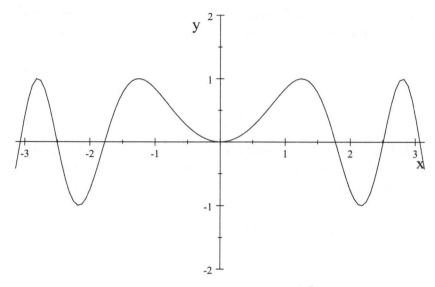

Figure 8.3 Graph of $y = \sin(x^2)$.

8.3.2 Problem: Derivative of $\sin(x^2)$

Find the derivative of $y(x) = \sin(x^2)$.

Solution

By the definition of derivative

$$
\begin{aligned}
y'(x) &= \lim_{t \to x} \frac{\sin(t^2) - \sin(x^2)}{t - x} \\
&= \lim_{t \to x} \frac{\sin(t^2) - \sin(x^2)}{t^2 - x^2} \times \frac{t^2 - x^2}{t - x}.
\end{aligned}
$$

If we let $t^2 = p$ and $x^2 = q$ in the first fraction then as $t \to x$, $p \to q$. We leave the second fraction as is and we get

$$
\begin{aligned}
y'(x) &= \lim_{p \to q} \frac{\sin(p) - \sin(q)}{p - q} \times \lim_{t \to x} \frac{t^2 - x^2}{t - x} \\
&= \cos(q)2x = 2x\cos(x^2).
\end{aligned}
$$

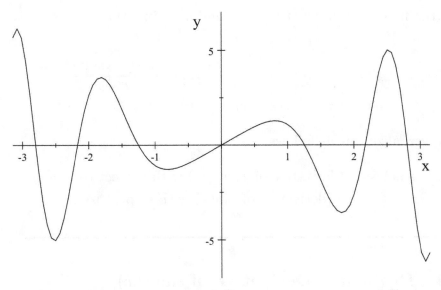

Figure 8.4 Graph of $2x\cos(x^2)$, the derivative of $\sin(x^2)$.

Think of $\sin(x^2)$ as being a composite function, one function inside another. The "square" is the inner function and the "sine" is the outer one. The above example suggests that the derivative of the composite function is the derivative of the outer function with respect to the inner one times the derivative of the inner function with respect to its variable.

8.3.3   Problem: Derivative of $f(\sin(x))$

In general, if f is any function defined and having a derivative on the interval $[-1, 1]$, and $y(x) = f(\sin(x))$, find $y'(x)$.

Solution

$$
\begin{aligned}
y'(x) &= \lim_{t \to x} \frac{f(\sin(t)) - f(\sin(x))}{t - x} \\
&= \lim_{t \to x} \frac{f(\sin(t)) - f(\sin(x))}{\sin(t) - \sin(x)} \times \frac{\sin(t) - \sin(x)}{t - x}.
\end{aligned}
$$

In the first fraction let $p = \sin(t)$ and $q = \sin(x)$, then as $t \to x$, $p \to q$, so

$$y'(x) = \lim_{p \to q} \frac{f(p) - f(q)}{p - q} \times \lim_{t \to x} \frac{\sin(t) - \sin(x)}{t - x}.$$
$$y'(x) = f'(q) \cos(x) = f'(\sin(x)) \cos(x).$$

We verbalize this as follows:

$$y'(x) = \text{(derivative of } f(\sin(x)) \text{ with respect to } \sin(x))$$
$$\times \text{(derivative of } \sin(x) \text{ with respect to } x). \qquad (8.4)$$

8.3.4   Problem: Derivative of $\sin(f(x))$

If $f(x)$ has a derivative on its domain, and $y(x) = \sin(f(x))$, find $y'(x)$.

Solution

$$y'(x) = \lim_{t \to x} \frac{\sin(f(t) - \sin(f(x))}{t - x}$$
$$y'(x) = \lim_{t \to x} \frac{\sin(f(t)) - \sin(f(x))}{f(t) - f(x)} \times \frac{f(t) - f(x)}{t - x}, \qquad (8.5)$$

now let $f(t) = p$, and $f(x) = q$, then $p \to q$ as $t \to x$, and (8.5) becomes

$$y'(x) = \lim_{u \to v} \frac{\sin(p) - \sin(q)}{p - q} \times \lim_{t \to x} \frac{f(t) - f(x)}{t - x},$$
$$y'(x) = \cos(q) \times f'(x) = \cos(f(x)) \times f'(x).$$

So, verbally,

$$y'(x) = \text{(derivative of } \sin(f(x)) \text{ with respect to } f(x))$$
$$\times \text{(derivative of } f(x) \text{ with respect to } x). \qquad (8.6)$$

8.4   Derivative notation

In the expression $y(x) = f(\sin(x))$, the inner function $\sin(x)$ acts as both a function, of x, and a variable for the outer function f.   When we say that we have found the derivative of f with respect to $\sin(x)$, we mean we have found the limit

$$\lim_{\sin(t)\to\sin(x)} \frac{f(\sin(t)) - f(\sin(x))}{\sin(t) - \sin(x)}.$$

We cannot denote this derivative as $(f(\sin(x)))'$ because the denominator is not $t - x$; it is $\sin(t) - \sin(x)$. Thus we say we have found the derivative of $f(\sin(x))$ with respect to its variable $\sin(x)$. One notation for this is

$$D_{\sin(x)} f(\sin(x)),$$

another is

$$\frac{df(\sin(x))}{d\sin(x)}.$$

Note that $(\sin(x))'$, the notation for the derivative of $\sin(x)$ with respect to its variable x may now be written as either:

$$D_x \sin(x)$$

or

$$\frac{d\sin(x)}{dx}.$$

Thus, if $y(x) = f(\sin(x))$, we can write

$$D_x y(x) = D_{\sin(x)} f(\sin(x)) D_x \sin(x)$$

or

$$\frac{dy(x)}{dx} = \frac{df(\sin(x))}{d\sin(x)} \times \frac{d\sin(x)}{dx}. \tag{8.7}$$

And for $y(x) = \sin(f(x))$,

$$D_x y(x) = D_{f(x)} \sin(f(x)) D_x f(x)$$

or

$$\frac{dy(x)}{dx} = \frac{d\sin(f(x))}{df(x)} \times \frac{df(x)}{dx}. \tag{8.8}$$

8.4.1 Problem: Find $\frac{d \sin^4(x)}{dx}$ and $\frac{d \sin(x^4)}{dx}$

Let $y_1(x)$ and $y_2(x)$ be the two functions

$$
\begin{aligned}
y_1(x) &= (\sin(x))^4, \text{and} \\
y_2(x) &= \sin(x^4).
\end{aligned}
$$

Use the derivative formula in Equations (8.7) and (8.8).

Solution

$$
\begin{aligned}
\frac{dy_1(x)}{dx} &= 4(\sin(x))^3 \times \cos(x). \\
\frac{dy_2(x)}{dx} &= \cos(x^4) \times 4x^3.
\end{aligned}
$$

Note that the derivative $\frac{dy(x)}{dx}$ may also be written as $\frac{d}{dx} y(x)$.

8.4.2 Problem: $\frac{d}{dx} \tan(\sin(x))$ and $\frac{d}{dx} \sin(\tan(x))$

1. If $y(x) = \tan(\sin(x))$, find $\frac{dy(x)}{dx}$.
2. If $z(x) = \sin(\tan(x))$, find $\frac{dz(x)}{dx}$.

Solutions

1.

$$
\begin{aligned}
\frac{dy(x)}{dx} &= \frac{d \tan(\sin(x))}{d \sin(x)} \times \frac{d \sin(x)}{dx} \\
&= \sec^2(\sin(x)) \cos(x).
\end{aligned}
$$

2.

$$
\begin{aligned}
\frac{dz(x)}{dx} &= \frac{d \sin(\tan(x))}{d \tan(x)} \times \frac{d \tan(x)}{dx} \\
&= \cos(\tan(x)) \times \sec^2(x).
\end{aligned}
$$

8.5   Derivative of composite functions $f(g(x))$

If $y(x)$ is a function of a function $f(g(x))$, assuming that the values $g(x)$ are in the domain of f, and that both functions have derivatives on their domains, then we want to find $y'(x)$.

8.5.1   Problem: Find $\frac{df(g(x))}{dx}$

If $y(x) = f(g(x))$, find $y'(x)$ and write it in the $\frac{dy(x)}{dx}$ notation.

Solution

By the definition of the derivative

$$y'(x) = \lim_{t \to x} \frac{f(g(t)) - f(g(x))}{t - x}.$$

Multiply the numerator and the denominator by $g(t) - g(x)$

$$y'(x) = \lim_{t \to x} \frac{f(g(t)) - f(g(x))}{g(t) - g(x)} \times \frac{g(t) - g(x)}{t - x}.$$

Now as $t \to x$, $g(t) \to g(x)$. In the first fraction let $g(t) = p$ and $g(x) = q$, then as $t \to x$, $p \to q$, so

$$y'(x) = \lim_{p \to q} \frac{f(p) - f(q)}{p - q} \times \lim_{t \to x} \frac{g(t) - g(x)}{t - x}, \qquad (8.9)$$

and by definition of the derivative

$$\lim_{p \to q} \frac{f(p) - f(q)}{p - q} = f'(q) = f'(g(x)),$$

$$\lim_{t \to x} \frac{g(t) - g(x)}{t - x} = g'(x).$$

So (8.9) becomes

$$y'(x) = f'(g(x))g'(x).$$

In the $D_x y$ notation this is

$$D_x y(x) = D_{g(x)} f(g(x)) \times D_x g(x),$$

or, in $\frac{dy}{dx}$ notation,

$$\frac{dy(x)}{dx} = \frac{df(g(x))}{dg(x)} \times \frac{dg(x)}{dx}. \tag{8.10}$$

Equation (8.10) is called the *chain rule* for the derivative of composite functions.

In the homework Problem 2, Exercise 8 you will get a chance to construct this rule for a longer chain of composite functions, $y(x) = f(g(h(x)))$. That is, you will prove the following formula.

$$\frac{dy(x)}{dx} = \frac{df(g(h(x)))}{dg(h(x))} \times \frac{dg(h(x))}{dh(x)} \times \frac{dh(x)}{dx}.$$

8.5.2 Problem: Find $\frac{d\sin(\cos(\tan(x)))}{dx}$

Use the chain rule to find the derivative of $\sin(\cos(\tan(x)))$ with respect to x.

Solution

$$\frac{d\sin(\cos(\tan(x)))}{dx} = [\cos(\cos(\tan(x)))] \times [-\sin(\tan(x))] \times \sec^2(x).$$

8.6 Differentiating a function

When we find the derivative of a function $f(x)$ we say we are *differentiating* the function. You can think of the word to *differentiate* as a verb and the *derivative* as a noun. When we differentiate a function, we are finding its derivative. We can differentiate a function that is defined in a given equation, without first solving the equation.

8.6.1 Problem Differentiate the equation $y^2 - 7x^3 + 100\sqrt{x} = 0$ with respect to x.

Differentiate the function $y^2(x) - 7x^3 + 100\sqrt{x}$ and differentiate 0.

Solution

$$\frac{d(y^2(x) - 7x^3 + 100\sqrt{x})}{dx} = 2y(x)\frac{dy(x)}{dx} - 21x^2 + \frac{50}{\sqrt{x}} = 0.$$

8.7   Exercise 8 HOMEWORK

1. Find the derivative of $(f(x))^{100}$.

2. $y(x) = f(g(h(x)))$, use the chain rule to find $\frac{dy(x)}{dx}$.

3. Find the derivative of the $\tan(\cos(\sin(x)))$.

4. If $y = \cos(x^3 + 1)$, find $\frac{dy}{dx}$.

5. If $y = \cos^3(ax)$, where a is a constant, find $y'(x)$.

6. If $y = (x^3 - 7x^2 + 10 + \sin(x))^{13}$, find $y'(x)$.

7. Differentiate $(x^7 - \frac{y(x)}{x})^{5/8}$.

8. Differentiate $\sqrt{\cos(x) + \sin(x)} - y(x) = 0$.

9. Differentiate $\sqrt{\frac{\sin(x)}{\cos(x)}}$.

10. If $y(x) = \sqrt{\sin(1/\cos(x^2))}$ find $\frac{dy}{dx}$.

11. Let $x(t) = \sin(t)$, and $y(x(t)) = u(t)$. If you know that $y'(x) = \sqrt{1 - x^2}$, can you find $u'(t)$? And if so, what is it?

12. If $y(x) = g[f(x)h(x) + k(u(x))]$, find $y'(x)$.

Chapter **9**

Implicit Differentiation of an Equation

9.1 Questions about implicit and explicit functions

- What are independent and dependent variables?

- When is a function explicitly defined?

- When is a function implicitly defined?

- How do we find the derivative of an implicitly defined function?

9.2 Independent and dependent variables

We will define what we mean by independent and dependent variables; this will facilitate the introduction of *explicit* and *implicit* functions. In stating these definitions, we assume F is a function whose domain is D and whose range is R.

9.2.1 Definition: INDEPENDENT VARIABLE

The statement that x *is an independent variable for* F means that x is an element in the domain, D.

In other words you select the variable x first then see what value y has that will make the pair (x, y) be in F.

9.2.2   Definition: DEPENDENT VARIABLE

The statement that y *is the dependent variable of* F means that y is an element of the range, R.

In other words, if the point (x, y) is in F, then the value of y depends on what value x has.

9.2.3   Definition: EXPLICIT EQUATION

When a function is defined by an equation in which the dependent variable, y, standing alone, is expressed in terms of the independent variable x such as

$$y = f(x) \tag{9.1}$$

then the function is *explicitly* defined.

In other words, we say y is an *explicit function* of x, and Equation (9.1) is called an *explicit equation.*

9.2.4   Example of an explicit function

$$y = \sqrt{100 - x^2}. \tag{9.2}$$

9.2.5   Definition: IMPLICIT EQUATION

If the dependent variable y is *not expressed* explicitly in terms of the independent variable, x such as

$$R(x, y) = 0, \tag{9.3}$$

then, we say the function is *implicitly* defined. The Equation (9.3) is called an *implicit equation* defining y as a function of x. Here the equation is not "solved for" y in the sense that y is not standing alone on one side of the equation with some expression in x standing on the other side.

9.2.6 Example of an implicit function

$$x^2 + y^2 - 100 = 0. \tag{9.4}$$

Some functions defined by Equation (9.4) are as follows:

$$
\begin{aligned}
y(x) &= \sqrt{100 - x^2}, \text{ if } -10 \le x \le 10 \text{ and} \\
y(x) &= -\sqrt{100 - x^2}, \text{ if } -10 \le x \le 10.
\end{aligned}
$$

But these are not the only functions that can be defined by Equation (9.4) There are infinitely many such functions. For example, another one is

$$
y(x) = \left\{
\begin{array}{ll}
\sqrt{100 - x^2} & \text{if } -10 \le x < 5 \\
-\sqrt{100 - x^2} & \text{if } 5 \le x \le 10
\end{array}
\right. .
$$

Whose graph is shown in Figure 9.1

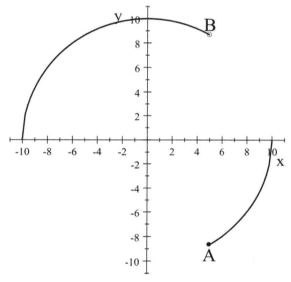

Figure 9.1 One of the functions defined by the equation $x^2 + y^2 - 100 = 0$.

The coordinates of all the points shown in Figure 9.1 satisfy Equation 9.4, and this figure is a function because no vertical line contains more than one point of it, but it is not a *continuous* function. We will discuss what it means for a function to be continuous in a later chapter. Right now, just notice that the point $A = (5, -8.66)$ is on the graph but if $x < 5$ and x approaches 5 the graph approaches $B = (5, 8.66)$ which is not on the graph.

9.2.7   Question: Are there any implicit equations that can't be solved for y?

Answer

Yes, for example, the following equation defines y as a function of x, but it cannot be solved explicitly for y in terms of x.

$$y + \sin(x + y) = 0. \tag{9.5}$$

9.2.8   Differentiation of an Implicit Function

Although we cannot always solve an implicit equation, we can still find the derivative of any of the functions it defines. Often such a derivative will not be obtained in terms of the independent variable, x alone, but will also have the dependent variable y as a part of it.

9.2.9   Example: Find the derivative of $x^2 + y^2 - 100 = 0$ with respect to x

Here we could solve for y and differentiate finding $\frac{dy}{dx}$.   But to illustrate implicit differentiation we will, first, write y as $y(x)$. Then we simply use the chain rule to differentiate $x^2 + y^2(x) - 100 = 0$ with respect to x.

$$\frac{d(x^2 + y^2(x) - 100)}{dx} = 2x + 2y(x)y'(x) = 0.$$

Solving for $y'(x)$ we get

$$y'(x) = \frac{-2x}{2y(x)} = -\frac{x}{y}.$$

This says that the slope at any point (x, y) of the graph of any function that satisfies the implicit Equation (9.4) will be $-\frac{x}{y}$, provided the graph has a slope at that point.

9.2.10 Example: Find the derivative of y if $y + sin(x + y) = 0$

If $y(x) + \sin(x + y(x)) = 0$, use implicit differentiation to find $\frac{dy}{dx}$.

$$\frac{d(y + \sin(x + y))}{dx} = \frac{dy}{dx} + \cos(x + y)(1 + \frac{dy}{dx}) = 0$$

$$= \frac{dy}{dx} + \cos(x + y) + \cos(x + y)\frac{dy}{dx} = 0. \qquad (9.6)$$

Now we solve Equation (9.6) for $\frac{dy}{dx}$, getting

$$\frac{dy(x)}{dx} = \frac{-\cos(x + y)}{1 + \cos(x + y)}.$$

9.2.11 Problem: Find $y'(x)$ if $y^2 + xy^3 = 10x^4$

Find the derivative of a function $y(x)$ that is defined by the given equation

$$y^2 + xy^3 = 10x^4. \qquad (9.7)$$

Solution

We could, first, try to solve the equation for y, as a function of x, then take its derivative, but it is easier to just differentiate the whole equation. If the two sides are equal then their derivatives must be equal, so.

$$\frac{d(y^2(x) + xy^3(x))}{dx} = \frac{d(10x^4)}{dx}$$

$$2y(x)y'(x) + x3y^2(x)y'(x) + y^3(x) = 40x^3,$$

$$y'(x) = \frac{40x^3 - y^3(x)}{2y(x) + x3y^2(x)}. \qquad (9.8)$$

In Equation (9.8) we have $y'(x)$, but we do not have it explicitly because it is in terms of $y(x)$ itself and not completely solved for in terms of x only.

9.3   Exercise 9 HOMEWORK

1. Assume that the equation

$$xy \sin(y) = y^2 + x + 1$$

   defines y as a function of x and that the derivative of $y(x)$ exists. Find $y'(x)$.

2. If $y^2 = \sin(\frac{1}{\cos^2(x)})$, find $\frac{dy}{dx}$.

3. If $x = \frac{y^2 + 3y + 1}{\sec(y)}$, find $y'(x)$.

4. If $\alpha(x) = \cos^{-1}(x)$, find the $\sin(\alpha(x))$ in terms of x,
   then find $\frac{d\sin(\alpha(x))}{dx}$.

5. If $y = \sqrt{\sin(\sec(x^2))}$, find $\frac{dy}{dx}$.

6. If $x^2 + 3xy + y^2 = \tan(y(x))$, find $\frac{dy}{dx}$.

7. If $\sqrt{x^2 + y^2} = 9$, find $\frac{dy}{dx}$.

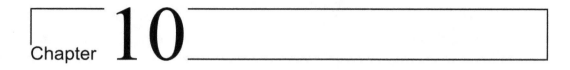

Chapter **10**

Inverse Functions

10.1 Questions about inverses

- The inverse of $\sin(x)$.

- Some trigonometric Identities.

- What is the derivative of the inverse $\sin(x)$?

- What are the derivatives of other inverse trigonometric functions?

10.2 The inverse of $\sin(x)$

Any number between 1 and -1 is the sine of some angle. For example $\frac{1}{2}$ is the $\sin(30°)$. In fact, $\frac{1}{2}$ is the sine of infinitely many other angles, $390°$, $750°$, and others every one of which is $30°$ plus some whole number multiple of $360°$. In the following definition, we will restrict ourselves to those angles that are between $-90°$ and $90°$.

10.2.1 Definition THE FUNCTION $\sin^{-1}(x)$.

If $-1 \leq x \leq 1$, then $\sin^{-1}(x)$ is the function $\alpha(x)$ such that $\sin(\alpha(x))$ is x. If this function, $\alpha(x)$, is measured in degrees, then α is between $-90°$ and $90°$. If it is measured in radians, then α is between $-\pi/2$ and $\pi/2$. See Figure 10.1. The equations defining $\sin^{-1}(x)$ are:

$$\sin^{-1}(x) = \alpha(x). \tag{10.1}$$
$$\sin(\sin^{-1}(x)) = \sin(\alpha(x)) = x. \tag{10.2}$$

In Figure 10.1 $\sin^{-1}(x) = \alpha$. It is the measure of $\angle ABC$.

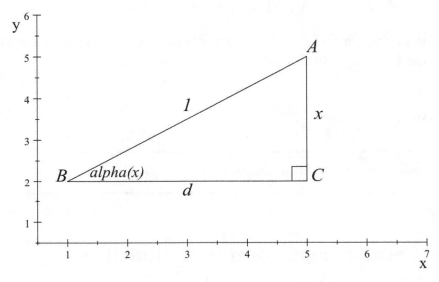

Figure 10.1 $\sin(\alpha) = x$, $d = \sqrt{1 - x^2}$.

10.2.2 Problem: Find $\tan(\sin^{-1}(x))$

Using Figure 10.1 find $\tan(\sin^{-1}(x))$ in terms of x.

Solution

$$\tan(\sin^{-1}(x)) = \tan(\alpha(x)) = \frac{x}{d} = \frac{x}{\sqrt{1-x^2}}.$$

10.2.3 Problem: What is $\cos(\sin^{-1}(x))$?

What is $\cos(\sin^{-1}(x))$?

Answer

$$\cos(\sin^{-1}(x)) = \cos(\alpha(x)) = \tfrac{d}{1} = \sqrt{1 - x^2}.$$

10.2.4   Puzzle: What is the $\sin^{-1}(\sin(x))$?

Since the inverse sine of x is a number whose sine is x, what is the number whose sine is the sine of x?

Answer

This is asking: what is $\sin^{-1}(\sin(x))$? The answer is x.

10.3   Some trigonometric identities

For any numbers n, α, and β,

1. $\sin(n\alpha) = 2\sin((n-1)\alpha)\cos(\alpha) - \sin((n-2)\alpha)$.
2. $\cos(n\alpha) = 2\cos((n-1)\alpha)\cos(\alpha) - \cos((n-2)\alpha)$.
3. $1 + \tan^2(\alpha) = \sec^2(\alpha)$.
4. $\cot^2(\alpha) + 1 = \csc^2(\alpha)$.

10.3.1   Problem: Prove Identity 1.

Prove $\sin(n\alpha) = 2\sin((n-1)\alpha)\cos(\alpha) - \sin((n-2)\alpha)$.

Solution

This is one way to do it.

$$
\begin{aligned}
2\sin(n\alpha - \alpha)\cos(\alpha) &= 2\left[\sin(n\alpha))\cos(\alpha) - \cos(n\alpha)\sin(\alpha)\right]\cos(\alpha) \\
&= 2\cos^2(\alpha)\sin(n\alpha) + \\
&\quad -2\sin(\alpha)\cos(\alpha)\cos(n\alpha). \tag{10.3} \\
\sin(n\alpha - 2\alpha) &= \sin(n\alpha)\cos(2\alpha) - \cos(n\alpha)\sin(2\alpha) \\
&= \sin(n\alpha)\left[\cos^2(\alpha) - \sin^2(\alpha)\right] + \\
&\quad -2\sin(\alpha)\cos(\alpha)\cos(n\alpha). \tag{10.4}
\end{aligned}
$$

Now subtracting Equation (10.4) from Equation (10.3) yields $\sin(n\alpha)$.

10.4   What is the derivative of $\sin^{-1}(x)$?

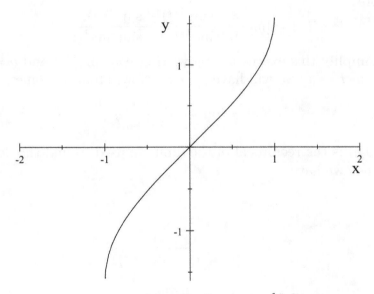

Figure 10.2 Graph of $y = \sin^{-1}(x)$.

10.4.1   Problem

If $\alpha(x) = \sin^{-1}(x)$, use the definition of derivative to find $\alpha'(x)$.

Solution

$$\alpha'(x) \;=\; \lim_{t\to x} \frac{\alpha(t) - \alpha(x)}{t - x}$$

$$\;=\; \lim_{t\to x} \frac{\sin^{-1}(t) - \sin^{-1}(x)}{t - x}.$$

But, by Equation (10.2),

$$\sin(\sin^{-1}(t)) = t$$

and

$$\sin(\sin^{-1}(x)) = x,$$

so

$$t - x = \sin(\sin^{-1}(t)) - \sin(\sin^{-1}(x)).$$

Therefore

$$\alpha'(x) = \lim_{t\to x} \frac{\sin^{-1}(t) - \sin^{-1}(x)}{\sin(\sin^{-1}(t)) - \sin(\sin^{-1}(x))}. \tag{10.5}$$

Let us simplify this expression by writing $p = \sin^{-1}(t)$ and $q = \sin^{-1}(x)$. Notice that as $t \to x$, we will have $p \to q$. Now, (10.5) becomes

$$\alpha'(x) = \lim_{p\to q} \frac{p - q}{\sin(p) - \sin(q)}. \tag{10.6}$$

But (10.6) is the reciprocal of the limit we used to find the derivative of $\sin(x)$. Hence we have

$$\alpha'(x) \;=\; \frac{1}{\cos(q)}$$

$$\;=\; \frac{1}{\cos(\sin^{-1}(x))},$$

$$\;=\; \frac{1}{\sqrt{1 - x^2}}.$$

Therefore, we have the formula

$$\alpha'(x) = \frac{d(\sin^{-1}(x))}{dx} = \frac{1}{\sqrt{1 - x^2}}. \tag{10.7}$$

10.5   Derivative by implicit differentiation

If $y(x) = \sin^{-1}(x)$, we can more easily find $y'(x)$ as follows.

$$
\begin{aligned}
y(x) &= \sin^{-1}(x) \\
\sin(y(x)) &= x, \\
\sin(y(x)) - x &= 0, \\
\frac{d(\sin(y(x)) - x)}{dx} &= 0, \\
\cos(y(x))y'(x) - 1 &= 0, \\
y'(x) &= \frac{1}{\cos(y(x))} = \frac{1}{\sqrt{1 - x^2}}.
\end{aligned}
$$

10.5.1   Problem: Find $\frac{d(\tan^{-1}(x))}{dx}$

If $y(x) = \tan^{-1}(x)$, find $\frac{dy}{dx}$.

Solution

If $y(x) = \tan^{-1}(x)$, then $x = \tan(y(x))$, now by implicit differentiation

$$
\begin{aligned}
x &= \tan(y(x)) \\
\frac{dx}{dx} &= \frac{d\tan(y(x))}{dx}, \\
1 &= \sec^2(y(x))\frac{dy}{dx}. \\
\frac{dy}{dx} &= \frac{1}{\sec^2(y(x))} = \frac{1}{\sec^2(\tan^{-1}(x))}, \\
&= \frac{1}{1 + \tan^2(\tan^{-1}(x))} = \frac{1}{1 + x^2}.
\end{aligned}
$$

In Figure 10.3, the point $A = (x, \tan^{-1}(x))$, and the tangent line has slope $m = 1/(1 + x^2)$.

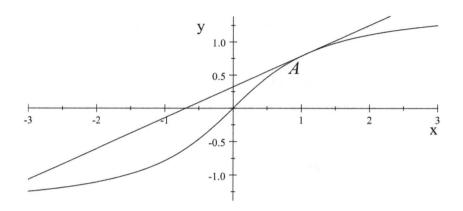

Figure 10.3 $\tan^{-1}(x)$ and the line with slope $\frac{1}{1+x^2}$ tangent at A.

10.6 Exercise 10 HOMEWORK

1. If $u(x) = \cos^{-1}(x)$, find $\frac{du}{dx}$.

2. If $z(t) = \sec^{-1}(t)$, rewrite this as $t = \sec(z(t))$ and differentiate both sides with respect to t in order to find $\frac{dz(t)}{dt}$.

3. If $y(x) = \sin^{-1}(\sqrt{x})$, find $y'(x)$.

4. If $y(x) = \tan^{-1}(f(x))$, where $f(x)$ has a derivative at x, find $\frac{dy(x)}{dx}$.

5. If $y(x) = (\sin^{-1}(x))^3$ find $\frac{dy(x)}{dx}$.

6. If f and g are inverse functions of each other, show that

$$\frac{df}{dg} = \frac{1}{\frac{dg}{df}}.$$

7. Solve the following problems

(a) If $a \neq 0$, and $u(x) = \tan^{-1}(\frac{x}{a})$, find $u'(x)$, and simplify as much as possible.

(b) Let h and k be two numbers. Based on your solution to part (a), find a function $u(t)$ such that

$$u'(t) = \frac{k}{k^2 + (t+h)^2}.$$

(c) Find a function $y(x)$ such that

$$y'(x) = \frac{1}{x^2 + x + 1}.$$

8. If $f(x) = g(h(x))$, prove the following version of the chain rule

$$\frac{dg}{dh} = \frac{\frac{df}{dx}}{\frac{dh}{dx}}.$$

Sometimes the equation $f(x) = g(h(x))$ is written as $g(x) = g(h(x))$. Although this does say that g is a function of x, it is inaccurate because the value of g at x is not the same as the value of g at $h(x)$. This practice is called "an abuse of notation" and it provides a symbolic short cut in certain expressions. In this case the resulting version of the chain rule is

$$\frac{dg}{dh} = \frac{\frac{dg}{dx}}{\frac{dh}{dx}}.$$

This form is appealing because it appears as if the elements dg, dh, dx (called *differentials*) are quantities that can be manipulated algebraically. Differentials can be and have been treated rigorously However, this is a subject that is more suitable for higher applied mathematics and engineering courses, and will not be discussed in this book.

Chapter 11

Second Derivative

11.1 Questions about second derivatives

- What happens to the slope of a graph near a maximum?

- The graph of a derivative.

- Speed and acceleration.

11.2 The slope near a relative maximum

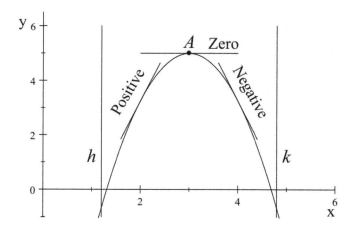

Figure 11.1 Slopes left and right of A.

11.2.1   Definition: RELATIVE MAXIMUM

A graph G has a relative maximum at a given point A if there exists a pair of vertical lines h and k with A between them such that no point of G between h and k is above the horizontal line through A.

11.2.2   Problem: What is the slope of a graph near a high point?

Suppose the point A is a relative maximum of a graph F.   Doesn't it seem reasonable to think that derivative will be positive at points to the left of and "close enough to" A, while points to the right of and "close enough to" A will be points at which the derivative will be negative?   Discuss this question.

Solution

An answer to this depends upon what is meant by the vague phrase "close enough".   It is possible to construct a graph that oscillates wildly to the left of a high point, so that, at some point very close to A on the left, the slope of the graph will be positive, while at some other point even closer to A on the left, the graph will have a negative slope.   We will see how to take care of this difficult problem when we discuss, in a later chapter, the *completeness axiom of the real number system* (also called the S_1S_2 axiom).

See Figure 11.1 where we have shown a piece of a graph between two vertical lines where A is the relative maximum.   In this figure we are assuming that at every point between the vertical line h and the point A, the graph has a positive slope.   Also, we assume that the graph has a negative slope at all of its points between A and the vertical line k.   This means the *slope, itself,* is decreasing, so *its* derivative is negative, for all points between h and k.

11.3   The graph of a derivative

Let (a, b) be a segment and F be the graph of a function $f(x)$ that has a derivative at each point of (a, b).   If $x_0 \in (a, b)$, the single number $f'(x_0)$ is the slope of F at a point $A = (x_0, f(x_0))$ For each element x of (a, b), let F'

be the set of all the ordered pairs: $(x, f'(x))$. That is F' is the graph whose ordinates are the derivatives of $f(x)$. This new graph is called the *derivative graph of F*. For example, let $f(x)$ be the function

$$f(x) = \frac{1}{5}x^5 + \frac{1}{4}x^4 - 7x + 2, \qquad (11.1)$$

then its derivative is

$$f'(x) = x^4 + x^3 - 7. \qquad (11.2)$$

Now let us draw the graphs F and the derivative graph of F, which we denote as F'. We do this by sketching, respectively, the graphs of the two functions

$$\begin{aligned} y_1 &= f(x), \text{ and} \\ y_2 &= f'(x). \end{aligned}$$

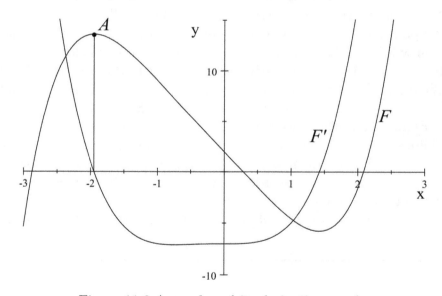

Figure 11.2 A graph and its derivative graph.

Notice that in Figure 11.2, the graph F has a relative maximum at the point A and the derivative graph F' is positive to the left of A and negative to the right of A.

11.4   The graph of a second derivative

Since $f'(x) = x^4 + x^3 - 7$ is also a function that has a derivative at each of its points, we can differentiate it getting

$$f''(x) = 4x^3 + 3x^2. \tag{11.3}$$

We can write the equation $y_3 = f''(x)$, and sketch its graph F'' as shown in Figure 11.3. This is the graph of the second derivative.

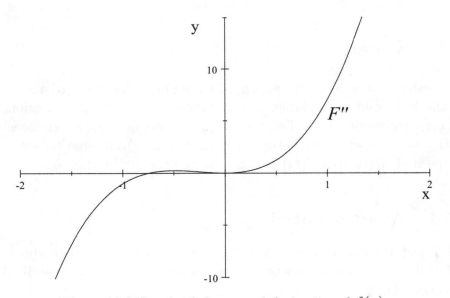

Figure 11.3 Graph of the second derivative of $f(x)$.

In Figure 11.4 we show the three graphs F, F', and F''.

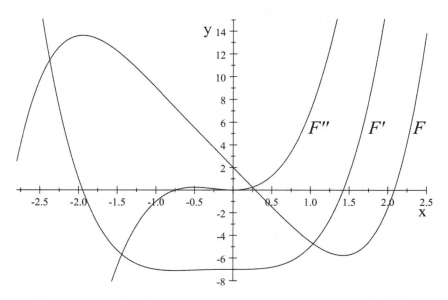

Figure 11.4 The first and second derivative graphs of $f(x)$.

11.5 Speed

In this section we will discuss an application of the first derivative. Suppose that the independent variable, t, represents *time* and the dependent variable, $y(t)$, represents some distance, area, volume, or some other measurable quantity that depends upon time, then for some given time t_0, we will see that the first derivative $y'(t_0)$ is the *velocity of* y at the time t_0.

11.5.1 Average speed

If an object travels some distance, x, in a given time, t, then the *average speed of the object* is the distance covered during the elapsed time divided by that time. That is, x/t.

11.5.2 Problem: Find the average speed

If an automobile travels 137 miles in 2 hours and 42 minutes, what is its average speed in miles per hour?

Solution

The time 2 hours and 42 minutes translates into $2\frac{42}{60}$ hours or 2.7 hours. The average speed is $\frac{137 \text{ miles}}{2.7 \text{ hours}} \approx 50.74$ m.p.h.

To say an object has a certain average speed, over a given interval, does not mean that it traveled at exactly that speed over the entire interval. It may travel at various speeds over several different subintervals of time.

11.5.3   Problem: Varying speed

An automobile starts on a trip in such a way that if t is the time (in hours) traveling, then its distance (in miles) from home, $x(t)$, is calculated by the following formula:

$$x(t) = -100\sqrt{t+1} + 85(t+1) + 15, \text{ where } t \geq 0. \qquad (11.4)$$

We set the time it leaves home to be at $t = 0$, so that $x(0) = -100 + 85 + 15 = 0$.

1. How far from home is the car 5 minutes after it leaves?

2. What is its average speed in m.p.h. over these first 5 minutes?

3. How far is the car from home one hour after it leaves and how far is it from home *one hour and five minutes* after it leaves?

4. What is the average speed over the 5 minute time interval from 1 hour to 1 hour and 5 minutes?

5. Find the average speed over the *one-minute* time interval between 2 hours to 2 hours and 1 minute.

6. How far did the automobile travel in 2 hours and 42 minutes? What was its average speed over the last 1 minute?

Solution

---

1. Five minutes translates to $\frac{5}{60}$ or 0.0833... of an hour, so

$$
\begin{aligned}
x(0.0833...) &\approx -100\sqrt{1.0833.} + 85(1.0833.) + 15 \\
&\approx 3 \text{ miles.}
\end{aligned}
$$

2. The average speed over the first five minutes is

$$
\frac{3 \text{ miles}}{0.0833 \text{ hour}} \approx 36 \text{ miles per hour.}
$$

3. This question is asking for $x(1)$ and $x(1.0833...)$

$$
\begin{aligned}
x(1) &= -100\sqrt{2} + 85 \times 2 + 15 \approx 43.58 \text{ miles, and} \\
x(1.0833) &\approx -100\sqrt{2.0833} + 85 \times 2.0833 + 15 \approx 47.74 \text{ miles.}
\end{aligned}
$$

4. We divide the distance traveled $x(1.0833) - x(1)$, by the time $1.0833 - 1$, getting

$$
\frac{47.74 - 43.58}{.0833} = \frac{4.16}{.0833} = 49.94 \text{ miles per hour.}
$$

5. The average speed over the one-minute interval from 2 hours to 2 hours and 1 minutes is:

$$
\frac{x(2 + \frac{1}{60}) - x(2)}{2 + \frac{1}{60} - 2} \approx \frac{97.73 - 96.79}{\frac{1}{60}} = 56.40 \text{ miles per hour.}
$$

6. Two hours and 42 minutes is 2.7 hours, and 2 hours and 41 minutes is 2.683... hours:

   The average speed over the last one minute is

$$
\frac{137 - 136}{2.7 - 2.683.} \approx \frac{1}{0.017} = 59 \text{ m.p.h.}
$$

11.5.4   Definitions: AVERAGE SPEED AND INSTAN-TANEOUS SPEED

Let X be a function such that the ordered pair $(t, x(t))$ belongs to X. Let h be a positive number and suppose that t_0 and $t_0 + h$ are in the domain of X, and that X is defined on the interval $[t_0, t_0 + h]$. If $x(t)$ is the distance x traveled in time t, then the **average speed** *of X over the interval* $[t_0, t_0 + h]$ is

$$\text{average speed} = \frac{x(t_0 + h) - x(t_0)}{h}.$$

If the size of the interval is allowed to approach zero by allowing $h \to 0$, then we can make the average speed become the **instantaneous speed** of X at t_0; when defined as follows:

$$r(t_0) = \lim_{h \to 0} \frac{x(t_0 + h) - x(t_0)}{h}.$$

11.5.5   Problem: Prove that the instantaneous speed is the derivative

Show that if g is a differentiable function on the interval $[t_0, t_0 + h]$, then the instantaneous speed, $r(t_0) = g'(t_0)$.

Solution

Let t_0 and t_1 be two numbers in the domain of g with $t_0 < t_1$. Now, let h be $t_1 - t_0$. Then $t_1 = t_0 + h$, and by the definition of derivatives

$$g'(t_0) = \lim_{t_1 \to t_0} \frac{g(t_1) - g(t_0)}{t_1 - t_0} = \lim_{h \to 0} \frac{g(t_0 + h) - g(t_0)}{h} = r(t_0).$$

11.6   Acceleration

Let $s(t)$ be the speed, in suitable units such as feet per second or miles per hour, of an object at any time, t. Now suppose that over a time interval

$[t_0, t_0 + h]$ we define the change in speed over a time interval $[t_0, t_0 + h]$ as

$$\text{Average acceleration} = \frac{s(t_0 + h) - s(t_0)}{h}.$$

We call this the *average acceleration* of the object over the given time interval.

11.6.1 Problem: Acceleration units

If $s(t)$ is speed in units feet per second, and t and $t + h$ are in seconds, what are the units for h and for the average acceleration?

Answer

Since both t and $t + h$ are in seconds, then h is in seconds. Since $s(t)$ and $s(t + h)$ are in feet per second, then the acceleration is in units of (feet per second) divided by seconds, or feet per second-squared.

$$\text{acceleration units} = \frac{\text{feet/second}}{\text{seconds}} = \text{feet/second}^2.$$

11.6.2 Instantaneous Acceleration

By letting $h \to 0$, we make the time interval decrease to zero and get the instantaneous acceleration at t_0. Thus

$$s'(t_0) = \lim_{h \to 0} \frac{s(t_0 + h) - s(t_0)}{h}.$$

But since $s(t)$ is speed, it is already a derivative, that is, $s(t) = x'(t)$. So acceleration is $x''(t)$.

11.7 Velocity, acceleration, and jerk.

In physics, we refer to the distance that an object moves in time t as displacement $x(t)$, how fast it moves as velocity $x'(t)$ and how fast the velocity changes as acceleration $x''(t)$. In engineering, we sometimes refer to how fast the acceleration changes (such as what a passenger in a speeding car might experience upon a sudden impact with a brick wall) as the "jerk". It is $x'''(t)$, the third derivative of $x(t)$.

$$
\begin{aligned}
x(t) &= \text{displacement.} \\
x'(t) &= \text{velocity.} \\
x''(t) &= \text{acceleration.} \\
x'''(t) &= \text{jerk.}
\end{aligned}
$$

11.8   Exercise 11 HOMEWORK

1. If $f(x) = (x - 1)x(x + 1)$

   (a) Find the first and second derivatives of $f(x)$.

   (b) Sketch the graphs of the three equations $y_1(x) = f(x)$, $y_2(x) = f'(x)$, and $y_3(x) = f''(x)$ in the same figure.

   (c) Find all points x, if any, where $f'(x) = 0$ and at each of these points find the value of f''.

2. If $y(x) = \sin(ax)$, where a is a constant, find $y'(x)$, $y''(x)$.

3. Show that $y(x) = \cos(3x)$ is a solution to the equation

$$
y''(x) + 9y(x) = 0.
$$

4. If $g(x) = \frac{1}{4}x^4 - 2x^3 + \frac{11}{2}x^2 - 6x + 10$

   (a) Find the three values x_1, x_2, and x_3 such that $g'(x_1) = 0$, $g'(x_2) = 0$, and $g'(x_3) = 0$.

   (b) Find the sign ($\pm$) of the second derivatives $g''(x_1)$, $g''(x_2)$, and $g''(x_3)$.

   (c) Identify the relative maxima and relative minima of the graph of $y = g(x)$.

5. Suppose that the "Shinkansen", the Japanese bullet train traveling from Kyoto to Tokyo, satisfies the following velocity equation, in Km. per hour.

$$v(t) = \begin{cases} 280t^{1/2} & \text{if } 0 < t < 1 \\ 280 + 0.1\left(16 - (t-2)^4\right)^{1/4} & \text{if } 1 < t < 3, \\ 280(4-t)^{1/2} & \text{if } 3 < t < 4. \end{cases}$$

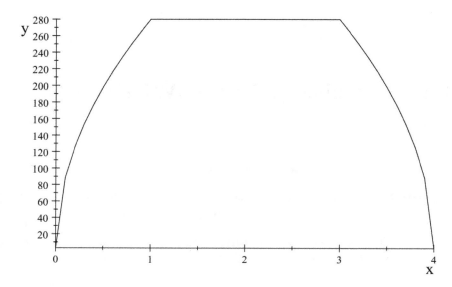

Figure 11.5 velocity of the "Shinkansen" bullet train in km/hour.

(a) Find the velocity at times $t = 0$, $t = 1/2$ hour $t = 2\frac{1}{2}$ hours, and at time $t = 3$ hours and 50 minutes.

(b) Find the acceleration at $t = 2\frac{1}{2}$ hours.

(c) Find the acceleration at $t = 3$ hours and 50 minutes. Why is it negative?

(d) Find the velocity and the acceleration $t = 3 : 59 : 30$, that is, at $1/2$ minute before the 4th hour.

(e) Show that as $t \to 1$ from the left, the acceleration is not equal to the acceleration as $t \to 1$ from the right. This ambiguity means that the acceleration at $t = 1$ cannot be defined.

6. Find the rectangle of maximum area that can be inscribed in the equilateral triangle $\triangle ABC$ whose sides are of length 2.

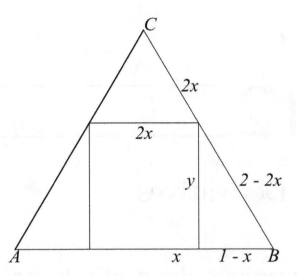

Figure 11.6 Rectangle inscribed in an equilateral triangle.

7. Find two functions, $g(x)$ and $h(x)$ such that

  (a) The graph of $y = g(x)$, has a relative maximum at a point x_0 and both $g'(x_0) = 0$ and $g''(x_0) = 0$.

  (b) The graph of $y = h(x)$, has a relative minimum at a point x_0 and both $h'(x_0) = 0$ and $h''(x_0) = 0$.

Chapter 12

Higher Derivatives

12.1 Questions about nth derivatives

- Definition of the *nth* derivative for any integer $n > 2$.

- Formulas for *nth* derivative of a product.

- Functions which do not have higher derivatives.

12.1.1 Definitions of higher derivatives

If $y = f(x)$, then we may denote the first and second derivatives as

$$y'(x) = \frac{dy(x)}{dx} \text{ and}$$
$$y''(x) = \frac{d^2y(x)}{dx^2}.$$

Similarly, the third derivative notation is

$$y'''(x) \text{ or } \frac{d}{dx}\frac{d^2y(x)}{dx^2} \text{ or } \frac{d^3y(x)}{dx^3}.$$

In general, if for any positive integer n, the *nth* derivative is denoted by

$$\frac{d^ny(x)}{dx^n},$$

117

then the $(n + 1)$ st derivative is

$$\frac{d^{n+1}y(x)}{dx^{n+1}} = \frac{d}{dx}\frac{d^n y(x)}{dx^n}.$$ (12.1)

The function $y(x)$, itself, is sometimes called the $0th$ derivative of $y(x)$, thus

$$\frac{d^0 y(x)}{dx^0} = y(x).$$ (12.2)

12.1.2   Problems

1. Let n be a positive integer and $y(x) = x^n$. Find

$$\frac{d^n y(x)}{dx^n}.$$

2. If $y(x) = (x + 1)\cos(2x) + x^3 + x^2 + x$. For $n = 1, 2$, and 3, find

$$\frac{d^n y(x)}{dx^n}.$$

3. If $y(x) = \cos^{-1}(x)$, find

$$\frac{d^2 y(x)}{dx^2}.$$

Solutions

_____

1. The nth derivative of x^n is $n!$

2. The first three derivatives of $(x + 1)\cos(2x) + x^3 + x^2 + x$ are

   (a) For $n = 1$,

   $$\frac{dy(x)}{dx} = \cos(2x) - 2(x + 1)\sin(2x) + 3x^2 + 2x + 1.$$

   (b) For $n = 2$,

   $$\frac{d^2 y(x)}{dx^2} = -4\sin(2x) - 4(x + 1)\cos(2x) + 6x + 2.$$

(c) For $n = 3$,

$$\frac{d^3 y(x)}{dx^3} = -12\cos(2x) + 8(x+1)\sin(2x) + 6.$$

3. $\frac{d^2}{dx^2}\cos^{-1}(x) = -x(1-x^2)^{-3/2}$, because

$$\frac{d}{dx}\cos^{-1}(x) = \frac{-1}{\sqrt{1-x^2}}$$
$$= -(1-x^2)^{-1/2}.$$

So

$$\frac{d^2}{dx^2}\cos^{-1}(x) = -\frac{d}{dx}(1-x^2)^{-1/2}$$
$$= -x(1-x^2)^{-3/2}.$$

12.1.3   Problem: Find the second derivative of a product

Let $g(x)$ and $h(x)$ be differentiable functions on the same domain, and if $y(x) = g(x)h(x)$, show that

$$\frac{d^2 y(x)}{dx^2} = g(x)\frac{d^2 h(x)}{dx^2} + 2\frac{dg(x)}{dx}\frac{dh(x)}{dx} + \frac{d^2 g(x)}{dx^2}h(x). \qquad (12.3)$$

Solution

When $y(x) = g(x)h(x)$, then by the product rule for derivatives

$$\frac{dy(x)}{dx} = \frac{d}{dx}(g(x)h(x)) = g(x)\frac{dh(x)}{dx} + h(x)\frac{dg(x)}{dx}.$$

Now by a repeated use of the product rule on these two terms, we get

$$\frac{d^2 y(x)}{dx^2} = \frac{d}{dx}\left[g(x)\frac{dh(x)}{dx} + h(x)\frac{dg(x)}{dx}\right]$$
$$= g(x)\frac{d^2 h(x)}{dx^2} + \frac{dh(x)}{dx}\frac{dg(x)}{dx} + \frac{dg(x)}{dx}\frac{dh(x)}{dx} + h(x)\frac{d^2 g(x)}{dx^2}.$$

This reduces to Equation (12.3).

12.2   Functions with no second derivative

In Exercise 11, we encountered a function (The Shinkansen velocity function) which failed to have a derivative at $t = 1$, although we could compute its derivative to the left of 1 and to the right of 1.   As we shall see later, this is an example of a function whose derivative is not continuous at $t = 1$.   In order for a function to have a derivative at a given point $P = (p, y(p))$, it must be true that the derivative $y'(t)$ has the *same* limit as $t \to p$, from either the left or the right.

12.3   Exercise 12 HOMEWORK

1. If $f(x) = x^4 + x^3 + x^2 + x + 1$

   (a) Find $f'(x)$ and $f''(x)$, the first and second derivatives of f.

   (b) Now find the first and second derivatives of f''.

2. Show that

$$\frac{d^3(g(x)h(x))}{dx^3} = g(x)\frac{d^3h(x)}{dx^3} + 3\frac{dg(x)}{dx}\frac{d^2h(x)}{dx^2}$$
$$+ 3\frac{d^2g(x)}{dx^2}\frac{dh(x)}{dx} + \frac{d^3g(x)}{dx^3}h(x).$$

3. Let n be a positive integer and $g(x)$ and $h(x)$ functions whose nth derivatives exist.   Express the nth derivative of the product $g(x)h(x)$ in terms of the derivatives of g and of h.

4. Find the derivatives

   (a) Suppose $y(x) = (x + 1)\sin(2x) + x^3 + x^2 + x$.

$$\text{Find } \frac{d^n y(x)}{dx^n} \text{ for } n = 1,\ 2,\ \text{and } 3.$$

   (b) If $y(x) = \sin^{-1}(x)$, then find $y''(x)$.

5. Let $y(x)$ be the function defined as follows

$$y(x) = \begin{cases} x & \text{if } 0 \le x \le 1 \\ -\frac{1}{2}x^2 + 2x - \frac{1}{2} & \text{if } 1 \le x \le 2 \end{cases}.$$

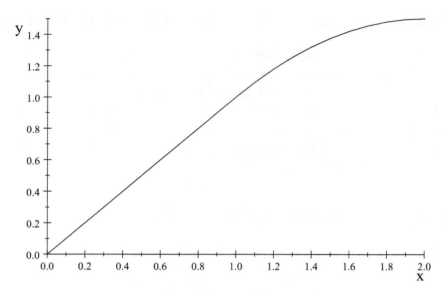

Figure 12.1 Graph of $y(x)$ Problem 5.

(a) Show that $y'(1)$ exists.

(b) Show that $y''(1)$ does not exist.

Chapter 13

Continuous Functions

13.1 Questions about continuity at a point

- How many functions can be defined by the same equation?

- What is meant by saying a function is continuous at a given point?

- When is a function discontinuous at a point?

13.2 Ambiguities with implicitly defined functions

The point set whose equation is

$$y^2 = x \tag{13.1}$$

is not a simple graph because there exists a vertical line that contains more than one point of it. (See Figure 13.1.)

123

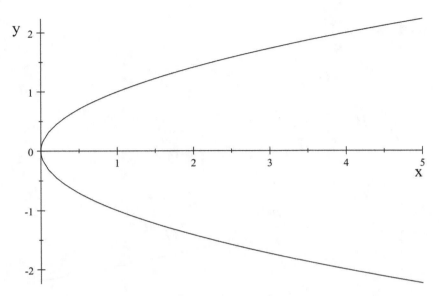

Figure 13.1 Graph of $y^2 = x$ on $[0, 5]$. Not a simple graph.

If we solve Equation 13.1 for y in terms of x, we get the ambiguous result that $y = \pm\sqrt{x}$. But by letting y be the positive square root of x for some values of x and letting y be the negative square root of x for other values of x, we can construct a function; in fact we can construct infinitely many functions by this method, one for each real number in the segment $(0, 5)$.

13.2.1   Example

Let x_0 be any real number in the segment $(0, 5)$ and define the function $y_{x_0}(x)$ as follows

$$y_{x_0}(x) = \begin{cases} \sqrt{x} & \text{if } 0 \le x \le x_0 \\ -\sqrt{x} & \text{if } x_0 < x \le 5 \end{cases}. \tag{13.2}$$

Example: If $x_0 = 1.58$, we get the graph, G, shown in Figure 13.2.

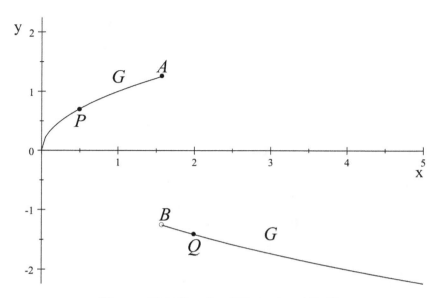

Figure 13.2 Graph of Equation (13.3).

The equation of G is

$$y_{1.58}(x) = \begin{cases} \sqrt{x} & \text{if } 0 < x \leq 1.58 \\ -\sqrt{x} & \text{if } 1.58 < x < 5 \end{cases} \tag{13.3}$$

The graph in Figure 13.2 is a simple graph on the segment $(0, 5)$. Here, G is a function whose domain is the segment $(0, 5)$. No vertical line between $x = 0$ and $x = 5$ contains more than one point of G.

By Equation (13.3), if $P = (x, y)$ is any point of G and $0 < x \leq 1.58$, then the ordinate of P is $+\sqrt{x}$. If $Q = (x, y)$ is a point of G with $1.58 < x < 5$, then the ordinate of Q is $y = -\sqrt{x}$. If A is the point of G whose abscissa is $x = 1.58$, then the ordinate of A is $y = \sqrt{1.58}$.

13.2.2   Problem

If B is the point of the graph in Figure 13.2, whose abscissa is $x = 1.58001$, show that the ordinate of B is $-\sqrt{1.58001} \approx -1.257$.

Solution

Since $x = 1.58001$, then $x > 1.58$, therefore by Equation (13.3), $y = -\sqrt{x} = -\sqrt{1.58001} \approx -1.257$.

13.2.3   Problem

Recall that on the graph G in Figure 13.2, A is the point with abscissa $x = 1.58$. Now let a and b be the two horizontal lines whose equations are $y = 1.6$ and $y = 1.14$. Show that the following statements are true:

1. The point A of G is between the horizontal lines a and b.

2. For *any* two vertical lines h and k, with A between them, there is some point of the graph G that is between h and k but is not between a and b. See Figure 13.3

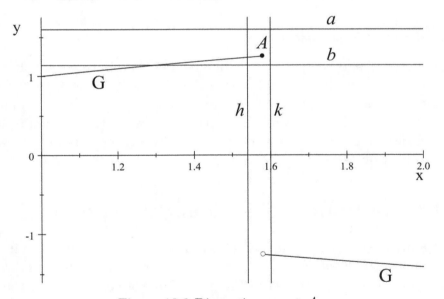

Figure 13.3 Discontinuous at A.

Solution

---

1. Since A has abscissa $x = 1.58$, and A is on G, then the ordinate of A is $y = \sqrt{1.58}$. We see that $1.14 < \sqrt{1.58} < 1.6$ because $(1.14)^2 < 1.58 < (1.6)^2$. Therefore A is between the horizontal lines a and b.

2. If h and k are any two vertical lines with A between them, then one of these vertical lines (say k) must be to the right of A. Let $x = r$ be the equation of k. The number r might be in the segment $(1.58, 5)$ or might even be greater than 5. Now, Let x_1 be any number such that the following two statements are true:

   (a) x_1 is less than 5 and

   (b) x_1 is between 1.58 and r.

Then x_1 is in the domain of G and between h and k. From Equation (13.3), $y_{1.58}(x_1) = -\sqrt{x_1}$, a negative number, hence less than $y = 1.14$. Therefore, $(x_1, y_{1.58}(x_1))$ is a point of G below the horizontal line b; not between a and b.

Summary of this solution

For the graph G and the point A belonging to G, we have shown that there exists two horizontal lines a and b with A between them such that if h and k are *any two* vertical lines with A between them, there is some point of G between the vertical lines h and k that is not between the horizontal lines a and b. As we shall see this is what we mean by saying G *is not continuous at the point* A on G.

13.3   Continuous functions at a point

We are going to state the definition of what is meant by saying that a function is *continuous at a given point*. You will see that from this definition, we have just proved that the function G in Figure 13.3 is not continuous at the point A.

13.3.1   Definition: CONTINUOUS AT A POINT

Let D be a number set and let G be a simple graph whose domain is D. The statement that G *is continuous at the point* A means:

1. The point A belongs to G, and

2. If a and b are any two horizontal lines with A between them, there exists two vertical lines h and k with A between them such that every point of G between h and k is also between a and b.

13.3.2   Problem

Let D be the interval $[-3, 3]$. If G is the graph of $y(x) = x$ for all $x \in D$, show that G is continuous at the point $A = (1, 1)$.

Solution

Let α, and β be any two horizontal lines with A between them. Let their respective equations be $y = a$ and $y = b$, $a > b$. Since the point A whose coordinates are $(1, 1)$ is between α and β, then $a > 1 > b$. See Figure 13.4. The graph G intersects α at the point (a, a), and β at the point (b, b). Now let h denote the vertical line $x = a$, and k the vertical line $x = b$. Every point of G between h and k is between α and β. Therefore G is continuous at A.

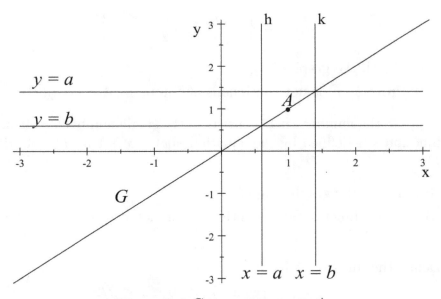

Figure 13.4 G is continuous at A.

13.3.3   Problem: What does "not continuous" mean?

What does it mean to say that a simple graph G is not continuous at a point A?

Solution

The statement that G is not continuous at the point A means that either

1. The point A does not belong to G or

2. There exist two horizontal lines a and b with A between them such that for *any* two vertical lines h and k with A between them, there is some point of G between h and k that is not between a and b.

13.4   Exercise 13 HOMEWORK

1. Define the function y_2 as follows

$$y_2(x) = \begin{cases} -\sqrt{x} & \text{if } 0 < x \le 2 \\ \sqrt{x} & \text{if } 2 < x \le 5 \end{cases}.$$

   (a) Sketch the graph of $y_2(x)$.

   (b) Prove $y_2(x)$ is not continuous at the point $A = (2, y_2(2))$.

2. Suppose the number set D consists of the single number $x = 0$, and the function s is defined by the equation $s(x) = x + 1$ for all x in domain D.

   (a) Sketch the graph of $s(x)$.

   (b) Is the function $s(x)$ continuous or not continuous at the point $A = (0, 1)$?

3. Define the function w as

$$w(x) = \begin{cases} \sin(x) & \text{if } 0 < x < \pi/2 \\ 0 & \text{if } \pi/2 \le x \end{cases}.$$

   Is w continuous at $x = \pi/2$?   Explain why or why not.

4. Define the function g as follows

$$g(x) = \begin{cases} \frac{x^2-1}{x-1} & \text{if } x \neq 1 \\ 2 & \text{if } x = 1 \end{cases}.$$

   (a) Sketch the graph of $g(x)$ for $-2 \leq x \leq 3$.

   (b) Is g continuous at the point $A = (1,2)$? Why or why not?

5. If D is the interval $[-3,3]$, and G is the graph of the equation $y = x^2$ on D, prove that g is continuous at the point $A = (1,1)$.

6. If a function $m(x)$ is defined as follows

$$m(x) = \begin{cases} 0 & \text{if } x \text{ is a rational number} \\ 1 & \text{if } x \text{ is an irrational number} \end{cases}.$$

   (a) What is $m(5/3)$?

   (b) What is $m(\pi)$?

   (c) Is m continuous at the point $x = 1$?

Chapter 14

Functions Continuous on Their Domains

14.1 Questions about continuity on a domain

- Rational and irrational numbers.

- Continuity on an interval.

- Continuity on a function's domain.

14.2 Rational and irrational domains

The x-axis is the set of all real numbers. Two important subsets of the real numbers are the *rational numbers* and the *irrational numbers*. The rational numbers consist of whole numbers and fractions such as $2/5, 75/13, -3/1000$, etc., in other words any number that can be written as the quotient of two whole numbers. The irrational numbers are ones that cannot be written as the quotient of two whole numbers. Some examples of irrational numbers are $\sqrt{2}$, π, $\sqrt[3]{31}$, $1/(\sqrt{5}-1)$, etc.

If a and b are any two numbers, there is a rational number between them, and if c and d are any two numbers, there is an irrational number between them. This means that there cannot be a segment (a, b) on the x-axis that consists *only* of irrational numbers, nor is it possible for a segment to consist of only rational numbers. It takes both rational and irrational numbers to

131

make up a complete segment. In this sense, we can say a segment is "full of holes," meaning that it is not completely accounted for by only the rational numbers, or by only the irrational numbers. However, a **domain** of some function might consist of only rational number or of only irrational numbers or of all the real numbers in a segment. These three situations can produce quite different results regarding the continuity of a function on its domain.

14.3   What is continuity on an interval?

In the previous chapter, we defined continuity at a point, now we ask what is meant by saying a function is continuous on its domain?

14.3.1   Definition: CONTINUITY ON AN INTERVAL

Let p and q be real numbers with $p < q$, and let D be the interval $[p, q]$. Let $g(x)$ be a function on D, and G the graph $y = g(x)$. If G is continuous at each of its points, then we say that *the function $g(x)$ is continuous on the interval D.*

14.3.2   Problem

What does it mean to say that a function $g(x)$ is not continuous on an interval D?

Solution

Given G is the graph of the equation $y = g(x)$. If $g(x)$ is not continuous on the interval D, then there exists some point A of G such that G is not continuous at A.

14.3.3   Definition: CONTINUITY ON A DOMAIN

If D is any number set and G is a simple graph on D, then the statement that G *is continuous on D* means that G is continuous at each of its points.

14.3.4 Problem

Let D be the set of all rational numbers between 0 and 1, (including 0 and 1). Let G be the graph of the function $y(x)$ defined as follows

$$y(x) = x, \text{ for all } x \in D.$$

1. Show that there exists a real number $t, 0 < t < 1$ such that the horizontal line α, with equation $y = t$ does not even intersect G.

2. Show that, never-the-less, $y(x)$ is continuous on D.

To solve this problem you may take for granted the fact that between any two real numbers there exist a rational number.

Solution

---

1. Since the domain D of G consists only of rational numbers, every point of G has rational coordinates $(\frac{m}{n}, \frac{m}{n})$ with the integers m and n having no common factors. Let α be the horizontal line with equation $y = \frac{1}{2}\sqrt{2}$. If α intersects G, at some point P on α then the coordinates of P are $(\frac{1}{2}\sqrt{2}, \frac{1}{2}\sqrt{2})$ but since G also contains P, its coordinates are the rational pair $(\frac{m}{n}, \frac{m}{n})$. This makes $\frac{1}{2}\sqrt{2}$ rational, which is false.

2. Let $A = (\frac{m}{n}, \frac{m}{n})$ be any point of G. To use the definition of continuity we can start by selecting any two horizontal lines *alpha* and *beta* with A between them. These two lines might not intersect G. Let the two lines have equations $y = a$ and $y = b$, with $a > b$, so that $a > \frac{m}{n} > b$. In order to insure that we are using horizontal lines that intersect G, let us now select two new horizontal lines *gamma* and *delta* between α and β, such that A is between γ and δ, and such that these two new lines have rational equations, say, $y = \frac{p}{q}$ and $y = \frac{s}{t}$, respectively. This is possible because between the two real numbers a and $\frac{m}{n}$ there is a rational number $\frac{p}{q}$. Similarly between $\frac{m}{n}$ and b, there is a rational number $\frac{s}{t}$.

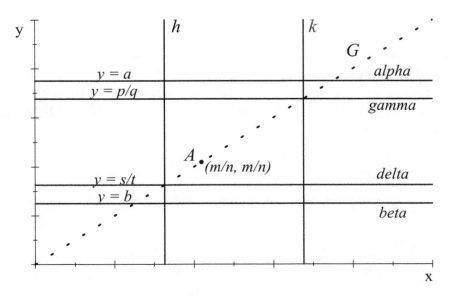

Figure 14.1 The graph G is continuous at A.

Now let h and k be the vertical lines with equations $x = p/q$ and $x = s/t$. Every point of G between h and k will be between γ and δ, hence between α and β, thus proving that G is continuous at A. Now since A was any point of G, we have proved that G is continuous at each of its points, so G is continuous on its domain.

14.4   Minimum points and continuity

If the domain D of a function, $f(x)$, is a segment and the function is continuous on D, and for some $x_0 \in D$, $f(x_0) < 0$, then f is negative in some segment containing x_0. This can be proved by solving the following problem.

14.4.1   Problem

Let a and b be real numbers, with $b > a$, and G be the graph of a continuous function $y = g(x)$, defined on the segment (a, b). If $g(x_0) < 0$ at some number x_0 in (a, b), show that there is some segment (c, d) containing x_0, such that $g(x) < 0$ for all $x \in (c, d)$.

Solution

If $g(x)$ is continuous on (a, b) and $g(x_0) < 0$, then for any two horizontal lines α and β with $A = (x_0, g(x_0))$ between them there exists two vertical lines h and k with A between them such that every point of the graph between h and k is also between α and β.

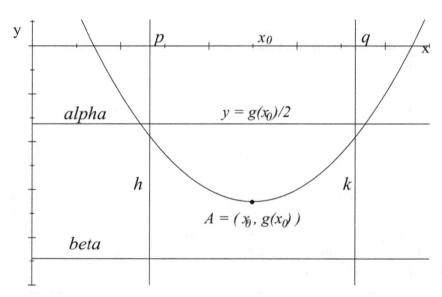

Figure 14.2. A graph where $g(x_0) < 0$.

Since $g(x_0) < 0$, then $g(x_0) < g(x_0)/2 < 0$. Why? Let us choose α to be the horizontal line whose equation is $y = g(x_0)/2$, and let β be any line below A. The line α is above A, therefore A is between α and β. There exists two vertical lines h and k with A between them such that every point of G between h and k is also between α and β.

If $(x, g(x))$ is any point of G between h and k, then $g(x) < 0$ because all such points are below the line $y = g(x_0)/2$. The vertical lines h and k intersect the x-axis in some points $(p, 0)$ and $(q, 0)$. Therefore, the segment (p, q) contains x_0, and $g(x) < g(x_0)/2 < 0$ for all x in (p, q). Let (c, d) be $(p, q) \cap (a, b)$. Then (c, d) contains x_0, a subset of (a, b), and for every point x in (c, d), $g(x) < 0$.

14.5   Exercise 14 HOMEWORK

1. Let D be the rationals between 0 and 1 (inclusive), and let $g(x)$ be the function defined as

$$g(x) = x^2, \text{ for all } x \in D.$$

   Prove that $g(x)$ is continuous at each of its points.

2. Let D be the interval $[0,1]$, and let $g(x)$ be the function defined for all $x \in D$ as

$$g(x) = \begin{cases} x^2 & \text{if} \quad x \text{ is rational} \\ 0 & \text{if} \quad x \text{ is irrational} \end{cases}.$$

   Prove that $g(x)$ is not continuous at any of its points, except at the origin.

3. Let a and b be real numbers, with $b > a$, and G be the graph of a continuous function $y = g(x)$, defined on the segment (a,b). If $g(x_0) > 0$ at some number x_0 in (a,b), show that there is some segment (b,d) containing x_0, such that $g(x) > 0$ for all $x \in (b,d)$.

4. Given that $y = m(x)$ is a twice differentiable function defined on a segment (a,b), $b > a$, and $m''(x)$ is continuous on (a,b) and there is a number $x_0 \in (a,b)$ such that $m'(x_0) = 0$ and $m''(x_0) > 0$, show

   (a) There is a subsegment $(c,d) \subset (a,b)$ containing x_0 such that $m''(x) > 0$ for all x in (c,d).

   (b) The graph of $y = m(x)$ has a relative minimum at $(x_0, m(x_0))$.

5. If $g(x)$ is a function that is twice differentiable and $g''(x)$ is continuous on its domain and x_0 is a number such that $g'(x_0) = 0$ and

   (a) if $g''(x_0) < 0$, prove that graph of $y = g(x)$ has a relative maximum at x_0.

   (b) if $g''(x_0) > 0$, prove that graph of $y = g(x)$ has a relative minimum at x_0.

   (c) if $g''(x_0) = 0$, prove, by examples, that the graph of $y = g(x)$ may have either an inflection point or a relative minimum or a relative maximum at x_0. This statement is sometimes written as follows. *The second derivative test* fails *or* is inconclusive.

Chapter **15**

Some Maximum - Minimum Problems

15.1 Questions about maxima and minima

- Nearest point problem.

- Ladder problems.

- Duality in maximum-minimum problems.

- Related rate problems.

15.2 Nearest point problem

If G is the graph of some equation $\mathcal{E}(x, y) = 0$, and $A = (x, y)$ is any point whose coordinates x, and y make the equation true, then the point A is on G; furthermore, all the points that are on G have coordinates that make the equation true. If $B = (p, q)$ is a point *not* on G then the coordinates p and q make the equation false. That is, $\mathcal{E}(p, q) \neq 0$. Indeed, all of points whose coordinates make the equation false are points that are not on G.

15.2.1 Example

Let G be the graph of the straight line equation $2x + 3y + 7 = 0$. Show that the point $A = (-5, 1)$ makes the equation true, but the point $B = (1, 2)$ makes the equation false.

137

Answer

Substitute in the coordinates of A and you get $2(-5)+3(1)+7 = -3+3 = 0$. The equation is true, so the point A is on the line. But, using the coordinates of the point B, we get $2(1) + 3(2) + 7 = 15$, not 0. So the equation is false. Therefore B is not on the line.

But how close is B to the line?

15.2.2   Problem: How far is the point, B from the line, L?

If B is the point $(1,2)$ and L is the line whose equation is $2x + 3y + 7 = 0$, find the distance from B to L.

Solution

By the distance from B to L we mean the *shortest* distance between B and a point of L. Consider the distance from $B = (1,2)$ to *any* point $X = (x,y)$ on L. Using the distance formula, we get

$$s = \sqrt{(x - 1)^2 + (y - 2)^2}.$$

Given that y depends on x, we write y as $y(x)$. Also, $s = s(x)$, therefore,

$$s(x) = \sqrt{(x - 1)^2 + (y(x) - 2)^2}.$$

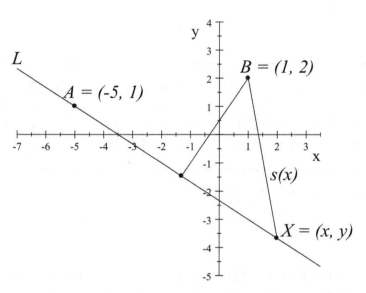

Figure 15.1 Distance from B to X.

The distance, $s(x)$, from B to X is *not* the shortest distance from B to L; it is simply the distance from B to a point on the line. See Figure 15.1. To find the point that provides the shortest distance, we need to find what value x_0 makes $s'(x) = 0$. Take the derivative of $s(x)$

$$s'(x) = \frac{2(x-1) + 2(y(x) - 2)y'(x)}{2\sqrt{(x-1)^2 + (y(x) - 2)^2}}.$$

This fraction is zero only if the numerator is zero. Hence, x_0 satisfies the following equation

$$(x_0 - 1) = (y_0 - 2)\frac{2}{3}, \text{where} \qquad (15.1)$$

$y_0 = y(x_0) = -(2x_0 + 7)/3$, so

$$x_0 - 1 = (-\frac{2x_0 + 7}{3} - 2)\frac{2}{3}.$$

Solving for x_0, we have $x_0 = -17/13$. Then $y_0 = -19/13$. The shortest distance from B to L is

$$s(\frac{-17}{13}) = \sqrt{(-\frac{17}{13} - 1)^2 + (-\frac{19}{13} - 2)^2} = \frac{15}{\sqrt{13}}.$$

This can also be written as

$$s(\frac{-17}{13}) = \frac{|2 \cdot 1 + 3 \cdot 2 + 7|}{\sqrt{2^2 + 3^2}} = \frac{15}{\sqrt{13}}.$$

In the homework problems, you will be asked to prove that the expression

$$s = \left| \frac{Ap + Bq + C}{A^2 + B^2} \right|, \tag{15.2}$$

is a *formula* for finding the distance from a point $P = (p, q)$ to a line $Ax + By + C = 0$. See Problem 1, Exercise 15.

15.3   Ladder problems

15.3.1   Problem: The paper cut problem

Suppose a point P is marked on a sheet of paper 8.5 inches by 11 inches, and this point is located 2.5 inches from the left edge and 4 inches from the bottom edge. The paper is to be cut along a straight line through the point P in such a way that the triangular piece formed by the two aforementioned edges and the straight line cut has the least possible area. Find the equation of that straight line.

Solution

We start by sketching the sheet of paper and showing a possible cut. See Figure 15.2.

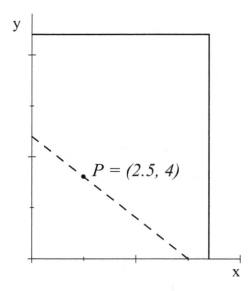

Figure 15.2 What is the triangle of minimum area?

Let us use the *intercept-form* of a straight line equation.

$$\frac{x}{a} + \frac{y}{b} = 1,$$

and let P be a point on the line

$$\frac{2.5}{a} + \frac{4}{b} \;=\; 1, \text{ or} \tag{15.3}$$

$$b \;=\; \frac{4a}{a - 2.5}. \tag{15.4}$$

The triangle is bound by the x-axis, y-axis and the first quadrant segment of the straight line. Its area is

$$A = \frac{1}{2}ab.$$

So from Equation (15.4), the area A, as a function of a, is

$$A(a) = \frac{1}{2}\frac{4a^2}{a - 2.5}.$$

We find $A'(a)$

$$A'(a) = \frac{2a^2 - 10a}{(a - 2.5)^2}.$$

Where is $A'(a) = 0$?   Only at either $a = 0$ or $a = 5$. Calculate $A''(x)$
and use the second derivative test for maxima or minima. $A''(0) < 0$, and
$A''(5) > 0$, therefore $a = 5$ furnishes the minimum.   From this, we have
$b = 8$. The area of the triangle is $\frac{1}{2}ab = 20$.

The cut that produces the triangle of minimum area is along the line
whose equation is

$$\frac{x}{5} + \frac{y}{8} = 1.$$

15.3.2   Problem: The ladder across the fence problem

A four foot tall fence runs parallel to a house and is two-and-a-half feet from
the house. Find the length of the shortest ladder that can reach the house
from outside the fence.

Solution

Let us draw two sketches representing the data in this problem

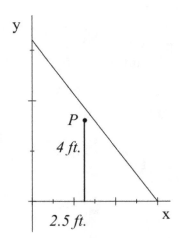

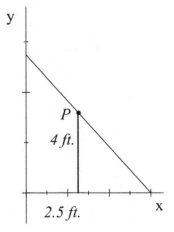

Figure 15.3a Ladder not   Figure 15.3b Ladder touching
touching the fence.   the fence.

Obviously, the ladder of minimum length must touch the fence. Otherwise, we could shorten the ladder until it did touch the fence. Again, using the *intercept-form* of a line passing through the point $P = (2.5, 4)$, we get the value of b from Equation (15.4), which we write expressing b as a function of a,

$$b(a) = \frac{4a}{a - 2.5}.$$ (15.5)

But this time the quantity we want to minimize is the length L as a function of a

$$L(a) = \sqrt{a^2 + b(a)^2}.$$

We use implicit differentiation to find $L'(a)$

$$L'(a) = \frac{a + b(a)b'(a)}{\sqrt{a^2 + b(a)^2}},$$

which is zero only when $a + bb'(a) = 0$, or

$$\begin{aligned} b'(a) &= \frac{-a}{b(a)} \\ &= \frac{-a}{\left(\frac{4a}{a-2.5}\right)}, \\ &= \frac{a - 2.5}{-4}. \end{aligned}$$ (15.6)

But if we compute $b'(a)$ directly from Equation (15.5), we get

$$b'(a) = \frac{-10}{(a - 2.5)^2}.$$

Hence

$$\frac{a - 2.5}{-4} = \frac{-10}{(a - 2.5)^2}.$$

So we must solve the equation

$$(a - 2.5)^3 = 40.$$

Therefore, $a \approx 5.92$ and $b \approx 6.92$. The ladder is $\sqrt{a^2 + b^2} \approx 9.1$ feet.

15.3.3   Problem: The ladder around the corner

If two hallways whose widths are 2.5 feet and 4 feet meet at a 90° angle, what is the length of the longest ladder that can be carried horizontally around the corner?

Solution

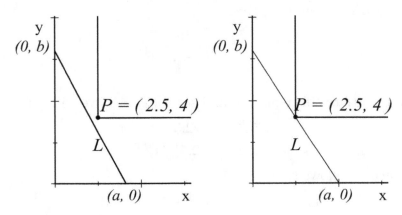

Figure 15.4a Ladder away   Figure 15.4b Ladder touching
from the corner.   the corner.

Analysis: First we consider a ladder of length L on the line $\frac{x}{a} + \frac{y}{b} - 1 = 0$, which we will rewrite as

$$bx + ay - ab = 0. \qquad (15.7)$$

The ladder passes into the second corridor as long as the line does not contain the point P, which has coordinates $(2.5, 4)$.   See Figure 15.4a and 15.54b. In other words, L is *not stuck* on the corner yet. If a is the x-*intercept*, then the y-*intercept*, b is $\sqrt{L^2 - a^2}$. For any a, the distance $s(a)$ from P to L is

$$s(a) = \left| \frac{2.5b + 4a - ab}{\sqrt{a^2 + b^2}} \right| . \qquad (15.8)$$

As long as $s(a) \neq 0$, the ladder is not touching the corner point and it will pass through. If we can find the length of the shortest ladder that contains the point P and touches both the x-axis and the y-axis, then we will have found the length of the longest ladder that can pass around the corner from one corridor to the next.

Note that there is no such thing as the longest ladder that contains the point P and touches both the x and y axes. So, just as in the ladder over the fence problem we want to find the value $a = a_0$, so that both $s(a_0) = 0$, and $L'(a_0) = 0$. First, setting $s(a) = 0$, we get

$$b = \frac{4a}{a - 2.5}.$$

Now since $L^2(a) = a^2 + b^2$, then

$$L^2 = a^2 + \frac{16a^2}{(a - 2.5)^2}.$$

If we differentiate this equation and set $L'(a) = 0$, we will get

$$(a - 2.5)^3 = (2.5)(4^2), \text{ or}$$
$$a - 2.5 = 2.5^{1/3}4^{2/3}$$

from which it follows that

$$a_0 = 2.5^{1/3}(2.5^{2/3} + 4^{2/3}), \text{ and}$$
$$b_0 = 4^{1/3}(2.5^{2/3} + 4^{2/3}).$$

Consequently, the length of the shortest ladder that can touch both axes and pass though the point P is

$$L = (2.5^{2/3} + 4^{2/3})^{3/2} \approx 9.1 \text{ feet},$$

which is the longest ladder that can go around the corner.

15.4 Duality in maximum minimum problems

15.4.1 Two classic problems

1. Given a fixed positive number P, find the rectangle of largest area, whose perimeter is P.

2. Given a fixed positive number A, find the rectangle of smallest perimeter whose area is A.

Solutions

---

1.   To find the maximum area for a given perimeter, let w and l be the width and length of any rectangle whose perimeter is $P = 2l + 2w$. Let A be the area, then

$$l = \frac{1}{2}P - w, \text{ and}$$
$$A = l \times w.$$

So the area, as a function of w, is

$$A(w) = (\frac{1}{2}P - w)w = \frac{Pw}{2} - w^2.$$

Now, $A'(w) = \frac{P}{2} - 2w$, which is zero when $w = P/4$; denote this value of w as w_0. Taking the second derivative, we get $A''(w_0) < 0$. This, with $A'(w_0) = 0$, tells us the A is maximum at $w_0 = P/4$. When we compute l, we get that l is also $P/4$. Therefore, the rectangle with a given perimeter has maximum area when the rectangle is a *square*.

2.   In this problem we want to find the minimum perimeter for a given area. Let w and l be the width and length of any rectangle whose area is A. Let P be the perimeter, then

$$P = 2l + 2w, \text{ and}$$
$$l = A/w.$$

So the perimeter, as a function of w is

$$P(w) = \frac{2A}{w} + 2w.$$

Now

$$P'(w) = \frac{-2A}{w^2} + 2.$$

And $P'(w) = 0$ when $w = \sqrt{A}$, making $l = \sqrt{A}$, hence the rectangle of a given area has a minimum perimeter when the rectangle is a *square*.

These two problems serve as an illustration of the principle of *duality*. One problem is a *dual* of another. Problem 1 asks for the maximum possible area of a rectangle under a given restriction on the perimeter and problem

2 asks for the minimum possible perimeter of a rectangle under a given restriction on its area. Both problems have the same answer, namely a square. Think of a way to argue that finding the shortest ladder over the fence is a dual of the problem of finding the longest ladder around the corner.

15.5 Related rate problems

Related rate problems ask you to compute the rate of change, per unit time, of a given quantity compared to the rate of change in a second quantity that is related to the first. For example, how fast is the volume of a sphere changing compared to how fast its radius is changing?

15.5.1 Problem: The inflating balloon

Suppose that a spherical balloon is being inflated in such a way that at some given time $t = 0$, the radius r is 7 centimeters and it is increasing at a rate of $\frac{1}{4}$ centimeter per second. How fast is the volume of the sphere increasing at that time?

Solution

The equation for the volume V of a sphere as a function of its radius r is

$$V = \frac{4}{3}\pi r^3.$$

If we assume that r is a function of time t, that is, $r = r(t)$, then the volume V will be a function of t, namely,

$$V(t) = \frac{4}{3}\pi r^3(t).$$

Differentiate and we get

$$V'(t) = 4\pi r^2(t)r'(t).$$

In this problem we are given that the radius at time 0 is 7 cm and the rate at which the radius is increasing is $\frac{1}{4}$ cm per second. That is $r(0) = 7$ and

$r'(0) = \frac{1}{4}$. Therefore the volume of the balloon at time $t = 0$ is increasing at the rate of $V'(0)$, that is

$$V'(0) = 4\pi r^2(0)r'(0)$$
$$= 4\pi \times 7^2 \times \frac{1}{4} = 49\pi \text{ cc/sec.} \approx 154 \text{ cc/sec.}$$

15.5.2   Problem: The sliding ladder

A ladder L feet long leans against a house and is sliding down the wall. At a specific time, $t = 0$, the foot of the ladder is a feet from the house and moving away from the house at a rate of v feet per second. How fast is the top of the ladder (that is the point $(0, b)$) moving downward?

Solution

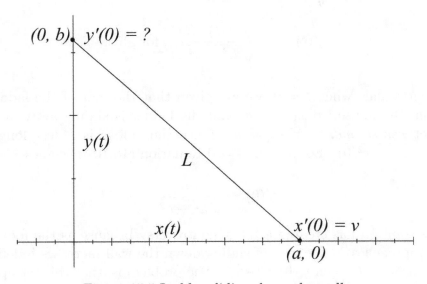

Figure 15.5 Ladder sliding down the wall.

At any time t let $x(t)$ be the distance of the foot of the ladder from the house, and $y(t)$ be the height of the top of the ladder. See Figure 15.5.

Then for all times t, the constant length L relates the distances $x(t)$ and $y(t)$ as follows

$$\sqrt{x(t)^2 + y^2(t)} = L. \tag{15.9}$$

If we differentiate Equation (15.9) implicitly, we get $x'(t)$ and $y'(t)$ as "related rates". Namely,

$$\frac{1}{2}(x^2(t) + y^2(t))^{-1/2}(2x(t)x'(t) + 2y(t)y'(t)) = 0,$$

$$\frac{x(t)x'(t) + y(t)y'(t)}{L} = 0.$$

In other words,

$$x(t)x'(t) + y(t)y'(t) = 0, \text{ for all } t.$$

Thus, the rate, $y'(t)$, at which the ladder is sliding down the wall at any time t is

$$y'(t) = \frac{-x(t)x'(t)}{y(t)}$$

$$y'(t) = \frac{-x(t)x'(t)}{\sqrt{L^2 - x^2(t)}}, \text{ for all } t. \tag{15.10}$$

In particular, when $t = 0$, we are given that the foot of the ladder is a feet from the house, so $x(0) = a$, and the ladder is sliding away at a rate of v feet *per second*, so $x'(0) = v$. Since the ladder is L feet long, then $y(0) = \sqrt{L^2 - x^2(0)}$. So, at time $t = 0$ Equation (15.10) becomes

$$y'(0) = \frac{-av}{\sqrt{L^2 - a^2}}.$$

This solution has created a bit of controversy because as the foot of the ladder approaches L, the rate of sliding down the wall increases indefinitely. That is, as $a \to L$, then $y'(0) = -\infty$. The problem is, that this set up is not really a good model for sliding ladders. This model does not include forces due to friction, nor forces bending the ladder and making it separate from the wall near the bottom.

15.6   Exercise 15 HOMEWORK

1. Let L be a line whose equation is $Ax+By+C = 0$, for any real numbers A, B, and C, with $A^2 + B^2 \neq 0$. Let $P = (p, q)$ be any point in the plane, show that the number s defined as

$$s = \left| \frac{Ap + Bq + C}{\sqrt{A^2 + B^2}} \right|,$$

   is the shortest distance from P to L.

2. If G is the graph whose equation is

$$y = 2\sqrt{x - 1} + 1, \text{ for } x \geq 1,$$

   and A is the point with coordinates $(4, 1)$, find the distance from A to G.

3. If G is the graph of the equation in Problem 2, and B is the point with coordinates $(2, 1)$, find the distance from B to G.

4. If a point $P = (p, q)$ is in the first quadrant of a coordinate system, find the line passing through P such that the triangle bound by that line and the two coordinate axes has the least area.

5. If a fence q feet tall is p feet away from a house, what is the length of the shortest ladder than can be placed across the fence and reach the house from a point on the ground outside the fence?

6. Two corridors meet at a right angle and the widths of the corridors are p feet and q feet, respectively. Let $P = (p, q)$ be the coordinates of the corner point of the two corridors.

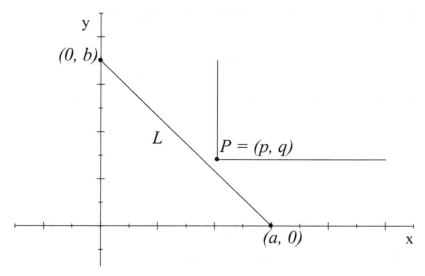

Figure 15.6 The y-intercept is $b = \sqrt{L^2 - a^2}$.

A ladder of length L is placed in such a way that the foot of the ladder is on the x-axis at a point $(a, 0)$ and the top of the ladder is at the point $(0, b) = (0, \sqrt{L^2 - a^2})$. See Figure 15.6.

(a) Find the length of the longest ladder that can be carried horizontally around the corner.

(b) Describe a solution to this problem as a dual of the ladder across the fence problem.

7. Two cars are traveling on straight roads one due east and the other due north, respectively. These two roads have a common intersection. At a specific time $t = 0$, the first car is 4 miles east of the intersection and driving east at 60 m.p.h. The second car is 5 miles north of the intersection and driving North at 40 m.p.h. At any time their *distance apart* is the hypotenuse of the triangle whose sides are on the two roads and whose lengths are the distances of each car from the intersection. Find how fast their distance apart is increasing at $t = 0$.

Chapter 16

Max-Min Problems in Geometry

16.1 Questions about areas, volumes and perimeters

- Is it required that a function always have a slope of zero at a maximum?

- Some Maximum Minimum problems in Geometry.

16.2 Can a graph have a maximum without a zero slope?

16.2.1 Problem: Find the highest point of $-\sqrt{9-x^2}$.

If G is the graph of $y = -\sqrt{9-x^2}$ on the interval $[0,3]$, find the lowest point of G and the highest point of G. See Figure 16.1

153

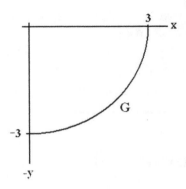

Figure 16.1 Graph of $y = -\sqrt{9 - x^2}$.

Solution

We can get the lowest point of G by setting the derivative $y'(x) = 0$; thus,

$$y'(x) = -\frac{1}{2}(9 - x^2)^{-1/2}(-2x) = \frac{x}{\sqrt{9 - x^2}},$$

which is zero only if $x = 0$. If we compute $y''(x)$, we will find that $y''(0) = \frac{1}{3}$. Thus, the graph is minimum at $x = 0$.

What about the maximum? We note that $y(x) \leq 0$, for all x, therefore the maximum value of y is *zero*, and this occurs at $x = 3$, where the derivative of y does not exist. This shows that a graph need not have zero slope at its maximum.

16.2.2   Problem

If G is the graph of the function $y = g(x)$, where $g(x) = 2x + 3$ for $x \in [0, 2]$, show that $g'(x)$ is never zero, and find the maximum and minimum values of G. What does this tell you about graphs that have their extreme points at their endpoints?

Solution

For all $x \in [0, 2]$, we have $g'(x) = 2$, so the derivative is never zero. The maximum value of G occurs at $x = 2$, since $g(2) = 7$, and for all other x in the domain $0 \le x < 2$ implies that $3 \le g(x) < 7$. Similarly, the minimum point of G is at $x = 0$, because $g(0) = 3$. A graph need not have a horizontal tangent at a maximum nor at a minimum that occurs at an endpoint.

16.3 Max-min problems in geometry

16.3.1 Problem: For a given area, find the cone of maximum volume.

An open right circular cone (one that does not include its circular base) and whose vertex is on a line perpendicular to the base has an area of 40 square inches. Find the dimensions that would make such a cone have the largest volume. See Figure 16.2.

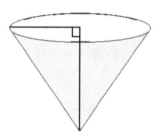

Figure 16.2 Open right circular cone.

Solution

If we cut the cone along a slant height, and flatten it out we will find that the surface is a circular sector, as shown in Figure 16.3. This major sector of the flattened cone has an arc length of $2\pi x$ being the circumference of the base of the cone.   The arc length is the central angle, θ, times its radius, h. Thus

$$\theta h = 2\pi x, \text{ or}$$
$$\theta = \frac{2\pi x}{h}.$$

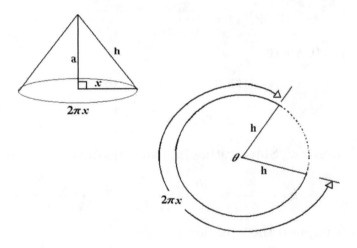

Figure 16.3 Right circular cone.

The area of the circular sector is one-half of the central angle times the square of the radius, h.

$$Area = \frac{1}{2}\theta h^2 = \pi x h.$$

We are given that the area is 40 in^2, so

$$\pi x h = 40, \text{ or}$$
$$h = \frac{40}{\pi x}.$$

The altitude $a = \sqrt{h^2 - x^2}$, and the volume of the cone is

$$V = \frac{1}{3}\pi x^2 a.$$

Putting a and h in terms of x, we get the following formula for the volume as a function of x,

$$V(x) = \frac{1}{3}x(1600 - \pi^2 x^4)^{1/2}.$$

Write this as

$$V(x) = \frac{1}{3}(1600x^2 - \pi^2 x^6)^{1/2},$$

and differentiate

$$V'(x) = \frac{\frac{1}{6}(3200x - 6\pi^2 x^5)}{\sqrt{1600x^2 - \pi^2 x^6}}.$$

Setting $V'(x) = 0$, yields

$$3200x = 6\pi^2 x^5, \text{ or}$$

$$x = \frac{1}{3^{1/4}}\sqrt{\frac{40}{\pi}}.$$

Call this solution x_0. Substituting into the equation for h, we have

$$h_0 = \frac{40}{\pi x_0} = \frac{40}{\pi}3^{1/4}\sqrt{\frac{\pi}{40}} = 3^{1/4}\sqrt{\frac{40}{\pi}}.$$

Then substituting into the equation for a,

$$a_0 = \sqrt{h_0^2 - x_0^2}$$

$$a_0 = \frac{4}{3^{1/4}}\sqrt{\frac{5}{\pi}}.$$

Thus the maximum volume for the cone with the given surface area is.

$$V_0 = \frac{1}{3}\pi x_0^2 a_0 = \frac{160}{3^{7/4}}\sqrt{\frac{5}{\pi}}.$$

16.4 Exercise 16 HOMEWORK

1. Find the minimum surface area of a right circular cylinder whose volume is 4 cubic inches. The surface area includes the top and bottom circular disks as well as the lateral surface. The volume of such a cylinder is $V = \pi r^2 h$. See Figure 16.4

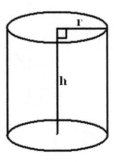

Figure 16.4 Cylinder with volume 4 in^3.

2. Dual of Problem 1: Find the cylinder of maximum volume whose total surface (including the lateral surface and the top and bottom disks) is a given fixed number, A.

3. Let G be the ellipse whose equation is

$$\frac{x^2}{16} + \frac{y^2}{9} = 1.$$

Find the foci of G. That is, find the two fixed points, F_1 and F_2, and a fixed number, k, *such that* if $P = (x, y)$ is any point of G, then the distance $\overline{PF_1}$ plus the distance $\overline{PF_2}$ is k.

4. Consider a square and a circle situated side by side as shown in Figure 16.5.   The base of the square and the diameter of the circle added together is 1 foot. Find the value of x so that the total area is as small as possible. Find a value of x, if any, so that the total area is as large as possible.

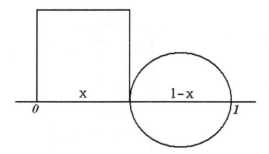

Figure 16.5 Find the maximum and minimum areas.

5. In *the quadrilateral ABCD* shown in Figure 16.6 there are three unit sides $\overline{AB}$, $\overline{BC}$, and $\overline{CD}$. The vertex angles at B and C are equal to each other. Find the height x that makes the quadrilateral have the largest possible area.

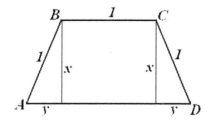

Figure 16.6 Find the maximum area.

Chapter 17

The Derivative of $\log_b(x)$

17.1 What is the calculus of logarithms?

- Find $\frac{d}{dx}\log_b(x)$ by definition of derivative.

- What is the limit of $(1 + \frac{1}{n})^n$?

- Intuition about fractions.

- An axiom about the positive integers.

- The number e.

- A natural way to simplify the derivative of a logarithm.

- The derivative of $\ln(u(x))$ and $e^{u(x)}$.

- Why is 1^{∞} an indeterminate form?

17.2 Finding the derivative of $y = \log_b(x)$

17.2.1 Problem: Find $\frac{d}{dx}\log_b(x)$.

If $b > 0$ and $b \neq 1$, and $y(x) = \log_b(x)$, use the definition of derivative to find $y'(x)$.

Solution

By the definition of derivative we have

$$y'(x) = \lim_{t \to x} \frac{\log_b(x) - \log_b(t)}{x - t}.$$

Let us use the law of logarithms that says $\log_b(M) - \log_b(N) = \log_b(\frac{M}{N})$, then

$$y'(x) = \lim_{t \to x} \frac{\log_b(\frac{x}{t})}{x - t}. \tag{17.1}$$

We are hoping to be able to write Equation (17.1) in some way that allows us to have $t \to x$, without creating the indeterminate form $\frac{0}{0}$. Avoiding the $\frac{0}{0}$ form is key to finding derivatives, but we have not achieved that goal yet because if we let $t = x$ in this equation we get

$$\frac{\log_b(\frac{x}{x})}{x - x} = \frac{0}{0}.$$

Keep trying. Now, by algebra, write Equation (17.1) as

$$y'(x) = \lim_{t \to x} \frac{1}{x - t} \log_b(\frac{x}{t}).$$

Then, using the law of logarithms that says $p\log_b(M) = \log_b(M^p)$, we will get

$$y'(x) = \lim_{t \to x} \log_b(\frac{x}{t})^{1/(x-t)}. \tag{17.2}$$

Simplify Equation (17.2) a bit by letting $x - t = h$, then $1/(x - t) = 1/h$. This also lets us write $x = t + h$, so $\frac{x}{t}$ is $\frac{t+h}{t} = 1 + \frac{h}{t}$. One more thing to notice is that since $x - t = h$, then $h \to 0$ as $t \to x$. So now, Equation (17.2) is

$$y'(x) = \lim_{h \to 0} \log_b(1 + \frac{h}{t})^{1/h}. \tag{17.3}$$

But have we made any progress? We still can't let $t = x$ because then we would have

$$\log_b(1^\infty).$$

As we will see in Section 17.9, the form 1^∞ is as bad an indeterminacy as $\frac{0}{0}$; so we keep trying.

Use the fact that $\log_b(A) = \frac{1}{t} \log_b(A^t)$ and rewrite Equation (17.3) as

$$y'(x) = \lim_{h \to 0} \frac{1}{t} \log_b (1 + \frac{h}{t})^{t/h}. \qquad (17.4)$$

We have $(1 + \frac{h}{t})^{t/h}$. Does it make any sense to raise a binomial to a fractional power? If we set $\frac{t}{h} = n$, then $\frac{h}{t} = \frac{1}{n}$, and

$$(1 + \frac{h}{t})^{t/h} = (1 + \frac{1}{n})^n. \qquad (17.5)$$

All we need to know is: what will we get if we let $n \to \infty$? This is the same as letting $h \to 0$, and $t \to x$.

At last! We have an expression that we may be able to evaluate – a binomial to a "whole" power. Let us *suspend* trying to find this derivative and take a closer look at the binomial expression

$$(1 + \frac{1}{n})^n. \qquad (17.6)$$

17.3   Limit of $(1 + \frac{1}{n})^n$ as $n \to \infty$.

Look at a few instances of $(1 + \frac{1}{n})^n$.

$$\text{If } n = 1, \text{ then } (1 + \frac{1}{n})^n = (1 + 1)^1 = 2,$$

$$\text{If } n = 2, \text{ then } (1 + \frac{1}{n})^n = (1 + 2 \times \frac{1}{2} + \frac{1}{4}) = 2\frac{1}{4},$$

$$\text{If } n = 3, \text{ then } (1 + \frac{1}{n})^n = (1 + 3 \times \frac{1}{3} + 3\frac{1}{9} + \frac{1}{27}) = 2\frac{10}{27}.$$

In general, for any positive integer n

$$\begin{aligned} (1 + \frac{1}{n})^n &= 1 + 1 + \frac{1}{2}(1 - \frac{1}{n}) + ... \\ &+ \frac{(1 - \frac{1}{n})...(1 - \frac{k-1}{n})}{k!} + ... + (\frac{1}{n})^n. \end{aligned} \qquad (17.7)$$

17.3.1   Problem: Find $(1 + \frac{1}{2000})^{2000}$.

Compute the first five terms of the series in Equation (17.7) when $n = 2000$.

Solution

$$(1 + \frac{1}{2000})^{2000} = 1 + 1 + \frac{1}{2}(1 - \frac{1}{2000}) + \frac{1}{3!}(1 - \frac{1}{2000})(1 - \frac{2}{2000})$$
$$+ \frac{1}{4!}(1 - \frac{1}{2000})(1 - \frac{2}{2000})(1 - \frac{3}{2000}).$$

This works out to about 2.708

17.4   Intuition about fractions

1. Can you think of some positive number y such that the fraction $\frac{1}{y}$ would be smaller than $\frac{3}{1000}$? Answer: One such number would be $y = 3000$, because the number $\frac{1}{3000}$ is smaller than $\frac{3}{3000}$ which is smaller than the number $\frac{3}{1000}$. Why?

2. What about a number y that would make $\frac{1}{y}$ be smaller than $\frac{10}{7000}$? The number $\frac{1}{8000} < \frac{10}{7000}$. Therefore $y = 8000$ would be one of many such numbers.

17.4.1   Problem: Make an assumption about small numbers.

Try to formulate an assumption about rational numbers (fractions) that illustrates, in general, the answers to the two questions above.

Solution

Your solution should say something like the following. *If x is any positive fraction then there is some number y big enough to make $\frac{1}{y}$ smaller than x.*

17.4.2 Question

If $m = 33$, can you find a positive integer n such that $\frac{m}{n} < 0.001$?

Answer

Yes, just pick n to be greater than 33000.

17.5 An axiom about the positive integers

We will now state an assumption about positive real numbers in order to complete the problem of finding the derivative of $\log_b(x)$. Before we state this axiom, here is another question which will help you understand its intuitive nature.

If $x = 500\pi$, can you find a positive integer, n, so that $\frac{x}{n} < 0.00005$? First, notice that $\pi < 4$, thus $500\pi < 2000$. We can change this problem so that it will be asking the same question about 2000 instead of 500π. Now simply pick n to be greater than 2000/0.00005. One, among infinitely many such numbers would be $n = 40,000,001$. Since $2000/40,000,000 = 0.00005$, then $2000/40,000,001 < 0.00005$. This means that $500\pi/40,000,001 < 0.00005$ also.

17.5.1 Problem

Make a assumption about positive real numbers that generalizes the above example.

Solution

Your solution should be a statement equivalent to the following one, which we will call *Axiom A*.

17.5.2   Axiom A

If x is any positive number, and δ is any positive number then there exists an integer n such that $\frac{x}{n} < \delta$.

17.5.3   Question: Limit of $\frac{1}{n}$

What does it mean to say that zero is the limit of $\frac{1}{n}$ as $n \to \infty$?

Answer

The statement: $\lim_{n\to\infty} \frac{1}{n} = 0$, means that if h is any horizontal line above the x-axis, then there is some positive integer N such that for every integer $n > N$, the point $(n, \frac{1}{n})$ is below h. In other words, if we let δ be any positive number and h be the horizontal line $y = \delta$, then there is some positive integer N, such that $\frac{1}{n} < \delta$, for all integers $n > N$.

17.5.4   Problem: What is the limit of $\frac{m}{n}$ for a fixed m as $n \to \infty$?

If m is any positive integer and for each positive integer n, a function F_m is defined as $F_m(n) = \frac{m}{n}$, prove

$$\lim_{n\to\infty} F_m(n) = 0.$$

Solution

If we can show that for any positive number δ we can always find an integer N large enough to make $|F_m(n)| < \delta$, for all integers, $n > N$, then we have proved that $F_m(n) \to 0$, as $n \to \infty$. Just use the axiom; given any positive number δ, then by Axiom A let N, be any positive integer such that $\frac{m}{N} < \delta$, then for all $n > N$, we have $\frac{m}{n} < \frac{m}{N}$. Therefore, $\frac{m}{n} < \delta$, hence $|F_m(n)| < \delta$ for all $n > N$.

17.6   The number e

17.6.1   Problem

Use Equation (17.7) to prove

$$\lim_{n\to\infty}(1+\frac{1}{n})^n = 1 + 1 + \frac{1}{2!} + \frac{1}{3!} + ...\frac{1}{k!} + ... \qquad (17.8)$$

Solution

The right hand side of Equation (17.7) starts with 1 and has $n+1$ terms ending with $(\frac{1}{n})^n$. If we let $n \to \infty$, we will get a series that does not end.

$$1 + 1 + \frac{1}{2}(1-\frac{1}{n}) + \frac{1}{3!}(1-\frac{1}{n})(1-\frac{2}{n}) + ...$$
$$+\frac{(1-\frac{1}{n})...(1-\frac{k-1}{n})}{k!} + ... \qquad (17.9)$$

In the previous problem, we have shown that for each positive integer m, the function $F_m(n) = \frac{m}{n} \to 0$, as $n \to \infty$. Thus all of the factors $(1-\frac{1}{n})$, $(1-\frac{2}{n}), ...(1-\frac{m}{n})$ approach 1 as $n \to \infty$. This proves that the expression (17.9) becomes

$$1 + 1 + \frac{1}{2!} + \frac{1}{3!} + ... + \frac{1}{k!} + ...$$

The value for this series is a number denoted by the letter e, and is approximately 2.71828.... Thus, we may restate Equation (17.8) as

$$\lim_{n\to\infty}(1+\frac{1}{n})^n = e.$$

17.6.2   Re-starting the derivation of $\frac{d\log_b(x)}{dx}$

17.6.3   Problem: Finish finding a formula for the derivative of $\log_b(x)$.

Complete the derivation of $\frac{d\log_b(x)}{dx}$.

Solution

Let us return to the problem of finding $y'(x)$, where

$$y(x) = \log_b(x).$$

We repeat Equation (17.4)

$$y'(x) = \lim_{h \to 0} \frac{1}{t} \log_b (1 + \frac{h}{t})^{t/h},$$

with $\frac{t}{h} = n$, and $x = t + h$. Therefore, as $h \to 0$, both $t \to x$, and $n \to \infty$. We may rewrite this as

$$y'(x) = \lim_{n \to \infty} \frac{1}{t} \log_b (1 + \frac{1}{n})^n.$$

Because $\log_b(x)$ is a continuous function, we can bring the limit, as $n \to \infty$, inside the $\log_b$ to get

$$y'(x) = \lim_{t \to x} \frac{1}{t} \log_b \big(\lim_{n \to \infty} (1 + \frac{1}{n})^n \big).$$

Therefore, we have

$$y'(x) = \frac{1}{x} \log_b(e),$$

so, if $b > 0$ and $b \neq 1$, then

$$\frac{d \log_b(x)}{dx} = \frac{1}{x} \log_b(e). \qquad (17.10)$$

17.7 Simplifying the derivative formula

Since the number e is, itself, a positive number and not equal to 1, it can be used as the base of a logarithm. If we wanted to find the derivative of $\log_e(x)$, then by Equation (17.10), we would have

$$\frac{d \log_e(x)}{dx} = \frac{1}{x} \log_e(e).$$

But since for any base, b, $\log_b(b) = 1$, this becomes

$$\frac{d \log_e(x)}{dx} = \frac{1}{x}.$$ (17.11)

Thus, it seems natural to use e as the base for logarithms. In fact, $\log_e(x)$ is called the *natural* logarithm and it even has its own notation "ln" instead of "$\log_e$". In other words, Equation (17.11) may be written as

$$\frac{d \ln(x)}{dx} = \frac{1}{x}.$$ (17.12)

17.7.1 Problem

Use the chain rule to find $y'(x)$, if $y(x) = \ln(x^4 + 1)$.

Solution

If $y(x) = \ln(x^4 + 1)$, then

$$\frac{d \ln(x^4 + 1)}{dx} = \frac{d \ln(x^4 + 1)}{d(x^4 + 1)} \frac{d(x^4 + 1)}{dx},$$

$$= \frac{1}{(x^4 + 1)} 4x^3.$$

17.8   The derivative of $\ln(u(x))$ and of $e^{u(x)}$

If $u(x)$ is a differentiable function, we will use the chain rule to find

$$\frac{d}{dx} \ln(u(x)),$$

and

$$\frac{d}{dx} e^{u(x)}.$$

17.8.1   Problem: Find $\frac{d}{dx}\ln(u(x))$.

Solution

Here we have a function of a function; we differentiate by the chain rule.

$$
\begin{aligned}
\frac{d}{dx}\ln(u(x)) &= \frac{d}{du}\ln(u)\frac{d}{dx}u(x), \\
&= \frac{1}{u(x)}u'(x) = \frac{u'(x)}{u(x)}.
\end{aligned}
$$

17.8.2   Example

If $y(x) = \ln(\tan(x))$, find $y'(x)$.

Solution

This derivative, $y'(x)$, can be computed either as the derivative of $\tan(x)$ divided by $\tan(x)$

$$
y'(x) = \frac{\sec^2(x)}{\tan(x)} = \frac{1}{\sin(x)\cos(x)}
$$

or, the derivative of $y(x) = \ln(\sin(x)) - \ln(\cos(x))$,

$$
y'(x) = \frac{\cos(x)}{\sin(x)} - \frac{-\sin(x)}{\cos(x)} = \frac{1}{\sin(x)\cos(x)}.
$$

17.8.3   Problem: Find $\frac{d}{dx}e^{u(x)}$

Solution

We want to find $y'(x)$, given that $y(x) = e^{u(x)}$. Take the ln of both sides,

$$
\ln(y(x)) = u(x).
$$

Then differentiate implicitly, getting

$$\frac{d}{dx}\ln(y(x)) = \frac{d}{dx}u(x).$$

So,

$$\frac{y'(x)}{y(x)} = u'(x).$$

Solving for $y'(x)$, we have,

$$\begin{aligned} y'(x) &= y(x)u'(x), \text{ that is} \\ \frac{d}{dx}e^{u(x)} &= e^{u(x)}\frac{d}{dx}u(x). \end{aligned}$$

17.9 Why is 1^∞ an indeterminate form?

When we first tried to find the derivative of $\log_b(x)$ from the definition of derivative we ran into an expression that would have resulted in the form 1^∞. We claimed that this is an indeterminate form just as bad as the expression $\frac{0}{0}$. Why is 1^∞ an indeterminate form? The reason is that it is an ambiguous result. In other words, it could have many different values; it is, actually, an indication that you have tried to compute a limit without simplifying it enough to get rid of the ambiguity.

17.9.1 Example

Find

$$\lim_{x\to\infty}(1+\frac{2}{x})^x.$$

If you simply substitute ∞ in for x you will get 1^∞, a meaningless form. If we first re-write $(1+\frac{2}{x})^x$ as $\left((1+\frac{2}{x})^{x/2}\right)^2$ and then let $\frac{x}{2} = n$, making $\frac{2}{x} = \frac{1}{n}$, the resulting expression is $\left((1+\frac{1}{n})^n\right)^2$. Now, as $x \to \infty$, and $n \to \infty$ and we get

$$\lim_{n\to\infty}\left((1+\frac{1}{n})^n\right)^2 = e^2,$$

making 1^∞ seem to stand for e^2 this time.

Let us consider another example.

$$(1+\frac{1}{n})^{n\ln(7)}.$$

When you put in $n = \infty$, you get 1^∞, but if you first re-write this as

$$\left((1 + \frac{1}{n})^n \right)^{\ln(7)},$$

then the limit as $n \rightarrow \infty$ is $e^{\ln(7)} = 7$.   So, this time, 1^∞ appears to stand for 7.

17.10   Exercise 17 HOMEWORK

1. If $y(x) = 10^x$, find $\log(y(x))$ and show that

$$\frac{d\log(y(x))}{dx} = 1.$$

2. If $y(x) = \log_5(x^2 + x + 1)$, find $y'(x)$.

3. If $g(x)$ is a differentiable function such that $\log_7 (g(x))$ is defined, find $\frac{d\log_7(g(x))}{dx}$ in terms of $g'(x)$.

4. If $y(x) = 10^x$, find $y'(x)$.

5. Find $\frac{d(e^x)}{dx}$.

6. Find $\frac{dy}{dx}$ if $y(x) = \ln(e^{\sin(x^2)})$.

7. If $y(x) = \sin(\ln(e^{x^2}))$, find $y'(x)$.

8. If $y(x) = \log(\sin^{-1}(\log(x)))$, find $y'(x)$.

9. If $y(x) = \log \sqrt{1 - x^2}$, find $y'(x)$.

10. Find the derivatives

    (a) If $y(x) = \ln(\cos(x))$, find $y'(x)$.
    (b) If $y(x) = \ln(\sec(x))$, find $y'(x)$.

11. Show how you can get the number π, from the indeterminate form 1^∞.

12. Prove, that if a is any real number then $\frac{d}{dx}x^a = ax^{a-1}$.

13. If $b > 0$, and $b \neq 1$, and $y(x) = b^x$, find $y'(x)$.

14. If $y(x)$ is the infinite series

$$y(x) = 1 + x + \frac{x^2}{2!} + \frac{x^3}{3!} + \frac{x^4}{4!} + \ldots \frac{x^n}{n!} + \ldots,$$

find the derivative $y'(x)$.

Chapter 18

The Function $Sin(\frac{1}{x})$

18.1 Questions about $\sin(1/x)$

- Where are all the roots of $\sin(1/x)$?

- A discussion of continuity of $\sin(1/x)$.

- A discussion of the continuity of $x\sin(1/x)$.

- For what values of n does the function $x^n\sin(1/x)$ have a derivative at $x = 0$?

18.2 Graph of $\sin(1/x)$

Let G be the graph of the equation

$$g(x) = \begin{cases} \sin(1/x) & \text{if } x \neq 0 \\ 0 & \text{if } x = 0 \end{cases}. \tag{18.1}$$

18.2.1 Problem

Construct a table showing those values of x for which $g(x) = 0, -1$, or 1, and use this to sketch a graph of G on the interval $[-2, 2]$.

Solution

---

x	$2/\pi$	$1/\pi$	$2/3\pi$	$1/2\pi$	$2/5\pi$	$1/3\pi$	$2/7\pi$	$1/4\pi$	...
$1/x$	$\pi/2$	π	$3\pi/2$	2π	$5\pi/2$	3π	$7\pi/2$	4π	...
$\sin(1/x)$	1	0	-1	0	1	0	-1	0	...

Figure 18.1 Table of values

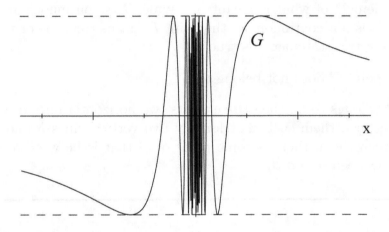

Figure 18.2 Graph of $\sin\left(\frac{1}{x}\right)$.

The dashed lines in Figure 18.2 are $y = 1$ and $y = -1$.

18.3   Continuity of $\sin(\frac{1}{x})$ at $(0,0)$

We want to see whether or not G is continuous at the point $P = (0,0)$.
Recall the definition that *a graph G is continuous at a point P* means that
both of the following two statements are true.

1. The point P belongs to G.

2. If α and β are any two horizontal lines with P between them then there
   exist two vertical lines h and k with P between them such that every
   point of G between h and k is also between α and β.

18.3.1 Question: When is a graph not continuous at a point?

What does it mean to say the graph G is not continuous at the point P?

Answer

When we want to say that *the graph G is not continuous at P*, we state the "bare denial" of what is meant by saying G *is* continuous at P. Thus, the graph G is not continuous at the point P means that one or the other of the following two statements is true.

1. The point P does not belong to G.

2. If P belongs to G, then there exists two horizontal lines α and β with P between them such that for any two vertical lines h and k with P between them, there is some point of G that is between h and k but not between α and β.

18.3.2 Problem: Is G continuous at the origin?

18.3.3 Guess

Make a guess as to whether or not the graph G in Figure 18.2 is continuous at the origin.

Solution

It is not. The next problem asks you to show that it is not.

18.3.4 Problem: Prove $\sin(1/x)$ is not continuous at $(0,0)$.

Show that the graph G defined in Equation (18.1), is not continuous at the origin.

Solution

For a graph to be continuous at a point, the first requirement is that the graph must contain the point, which this graph does because $g(0) = 0$, but it does not meet the second requirement. We show this now. Let us imagine that the graph in Figure 18.3 represents the graph of G near the origin. Actually, no matter how close we get to the origin, the graph would still look like Figure 18.2 So, an image such as the one in Figure 18.3, a "cartoon", is sometimes used to illustrate the essential elements of the discussion.

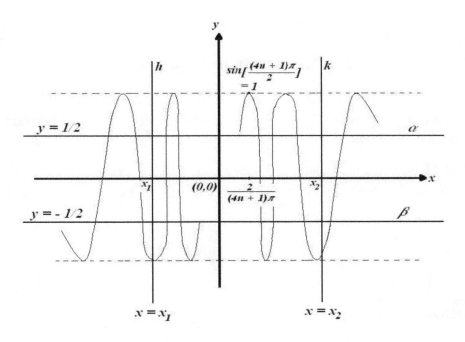

Figure 18.3

Let α and β be the horizontal lines whose equations are $y = \frac{1}{2}$ and $y = -\frac{1}{2}$, respectively. If h and k are *any two* vertical lines with equations $x = x_1$ and $x = x_2$, respectively, and with P between them, then either $x_1 < 0$ and $x_2 > 0$, or the other way around. Say $x_1 < 0 < x_2$. By **Axiom A** in Chapter 17, we know that there is some positive integer n such that $0 < \frac{1}{n} < x_2$, and since for any integer n, $(4n+1)\pi > n$, then $0 < \frac{1}{(4n+1)\pi} < \frac{1}{n}$, then $\frac{1}{(4n+1)\pi} < x_2$. By the definition of g, we know $g(\frac{2}{(4n+1)\pi}) = 1 > \frac{1}{2}$.

Therefore the point $(\frac{2}{(4n+1)\pi}, 1)$ is a point of G between h and k, and above α. Roughly, this means that no matter how close you bring the vertical line to the x-axis, the graph will always have points, "escaping" (not between α and β). In other words as $x \to 0$, y oscillates from $+1$ to -1, and does not approach 0. This proves that G is not continuous at $(0,0)$.

18.3.5 Problem: Is G continuous when $x \neq 0$?

What about the graph of $y = \sin(1/x)$ when $x \neq 0$? We will consider the case where x is positive.

Let G be the graph defined in Equation (18.1) and let $x_1 > 0$. If $P = (x_1, \sin(1/x_1))$, show that G is continuous at P.

Solution

Let α and β be horizontal lines with equations $y = a$ and $y = b$, respectively, and with P between them. See Figure 18.4. There exist vertical lines, h and k whose equations are $x = 1/\sin^{-1}(a)$, and $x = 1/\sin^{-1}(b)$ with P between them such that every point of G between these vertical lines is also between the two given horizontal lines. This proves that G is continuous at every point with a positive abscissa. A similar proof can be given for all $x < 0$.

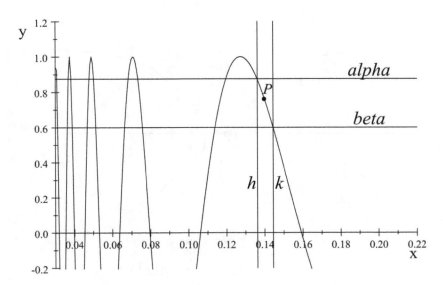

Figure 18.4 Is G continuous at P?

18.4   Continuity of $y = x \sin(\frac{1}{x})$

Let H be the graph of the following function $h(x)$.

$$h(x) = \begin{cases} x \sin(\frac{1}{x}) & \text{if } x \neq 0 \\ 0 & \text{if } x = 0 \end{cases}.$$ (18.2)

18.4.1   Problem

Sketch the graph of H.

Solution

A table of values for $h(x)$ can be inferred from Table 1. Thus, $h(2/\pi) = 2/\pi, h(1/\pi) = 0, h(2/(3\pi)) = -2/(3\pi), \dots$ We use these to sketch the graph in Figure 18.5 The dashed lines are the graphs of $y = x$ and $y = -x$.

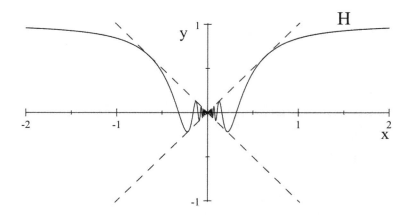

Figure 18.5 Graph of $x \sin(\frac{1}{x})$.

18.4.2   Problems

1. Answer the following questions about the graph H

   (a) Is H continuous at $x = 0$?

   (b) Is H differentiable at $x = 0$?

Solutions

We leave these two questions as homework problems.   See Problem 1, Exercise 18.

18.5   Differentiability of $x^2 \sin(\frac{1}{x})$ at $x = 0$.

We just saw that the function $\sin(\frac{1}{x})$ is not continuous at $(0,0)$, so we know that it does not have a derivative at that point. It does have a derivative at all of its other points.

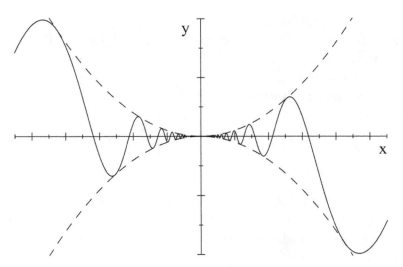

Figure 18.6 $y = x^2 \sin(1/x)$.

Now we want to find out whether or not the function $x^2 \sin(\frac{1}{x})$ has a derivative at $x = 0$. See the graph of this function in Figure 18.6. The dashed lines are $y = x^2$ and $y = -x^2$.

18.5.1   Problem

Let R be the graph of the function $r(x)$ defined by the equation

$$r(x) = \begin{cases} x^2 \sin(\frac{1}{x}) & \text{if} \quad x \in [-2, 2] \text{ but } x \neq 0 \\ 0 & \text{if} \quad\quad\quad x = 0 \end{cases}. \tag{18.3}$$

1. Show that R is continuous at $(0, 0)$.

2. From the rules for derivatives, show that $r'(x)$ exists for all $x \neq 0$.

3. Show that $r'(x)$ exists at $x = 0$.

Solution

---

1. Denote the point $(0, 0)$ by P. Take any two horizontal lines $y = a$ and $y = b$ with P between them. Either a or b is positive and the other is negative. Say $a > 0 > b$. Let c be the minimum of the absolute values

$|a|, |b|$. Then the lines whose equations are $y = c$ and $y = -c$ are two horizontal lines with P between them and these are also between the original two horizontal lines. These two lines intersect the graph of $y = x^2$ at the four points: $(-\sqrt{c}, c), (\sqrt{c}, c), (-\sqrt{c}, -\sqrt{c})$ and $(\sqrt{c}, c)$. Now since the sine function is never greater than one in absolute value, we can say that if $x \in [-\sqrt{c}, \sqrt{c}]$, then

$$-c \le x^2 \sin(\frac{1}{x}) \le c.$$

Therefore if h and k are the two vertical lines whose equations are $x = -\sqrt{c}$, and $x = \sqrt{c}$, then P is between h and k, and every point of R between h and k is also between $y = a$ and $y = b$. Hence R is continuous at P.

2. If $x \ne 0$, then by the rules for derivatives

$$r'(x) = 2x \sin(\frac{1}{x}) - \cos(\frac{1}{x}). \tag{18.4}$$

3. Note that $r(0) = 0$, and use the definition of derivative. To find $r'(0)$, we write

$$
\begin{aligned}
r'(0) &= \lim_{t \to 0} \frac{r(t) - r(0)}{t - 0} \\
&= \lim_{t \to 0} \frac{t^2 \sin(\frac{1}{t}) - 0}{t - 0}, \\
&= \lim_{t \to 0} t \sin(\frac{1}{t}). \tag{18.5}
\end{aligned}
$$

Now, we know that $|\sin(\frac{1}{t})| \le 1$, for all t, therefore $|t \sin(\frac{1}{t})| \le t$, thus the limit in Equation 18.5 is zero. Hence $r'(0) = 0$. We can summarize Equations (18.4) and (18.5) as

$$r'(x) = \begin{cases} 2x \sin(\frac{1}{x}) - \cos(\frac{1}{x}) & \text{if } x \ne 0 \\ 0 & \text{if } x = 0 \end{cases}. \tag{18.6}$$

18.6   Exercise 18 HOMEWORK

1. Sketch the graph of the function $h(x)$, defined as follows

$$h(x) = \begin{cases} x\sin(\frac{1}{x}) & \text{if } x \neq 0 \\ 0 & \text{if } x = 0 \end{cases}.$$

   (a) Show that $h(x)$ is continuous at $(0,0)$.

   (b) Show that $h(x)$ does not have a derivative at $(0,0)$, but it has a derivative at all other points.

2. Let a be any real number and F be the graph of a function $f(x)$ defined on the interval $[-2, 2]$ as follows

$$f(x) = \begin{cases} \sin(\frac{1}{x}) & \text{if } x \neq 0 \\ a & \text{if } x = 0 \end{cases}.$$

   Determine whether or not there could be any value of the variable a, that could make the graph F be continuous at $(0, a)$?   Why or why not?

3. Sketch the graph of

$$h(x) = \begin{cases} x^4 \sin(\frac{1}{x}) & \text{if } x \neq 0 \\ 0 & \text{if } x = 0 \end{cases}.$$

   (a) Show that $h(x)$ is continuous at $(0,0)$.

   (b) Show that $h(x)$ has a derivative at $(0,0)$ and at all other points.

4. Show that the derivative $r'(x)$ in Equation (18.6)

$$r'(x) = \begin{cases} 2x\sin(\frac{1}{x}) - \cos(\frac{1}{x}) & \text{if } x \neq 0 \\ 0 & \text{if } x = 0 \end{cases}$$

   is not continuous at $x = 0$. Hint: Can we possibly define $\cos(1/x)$ at $x = 0$ to make it continuous?

5. Show that if $0 < p < 1$, then the function, $f(x)$, defined by

$$f(x) = \begin{cases} x^p \sin(\frac{1}{x}) & \text{if } x \neq 0 \\ 0 & \text{if } x = 0 \end{cases}$$

   does not have a derivative at the origin.

6. Show that if $p > 1$, then $f(x)$, defined in Problem 5, does have a derivative at the origin.

7. Show that if $p > 2$, the derivative found in Problem 6 *is* continuous at the origin.

8. Show that if $1 < p < 2$, the derivative found in Problem 6 is not continuous at the origin.

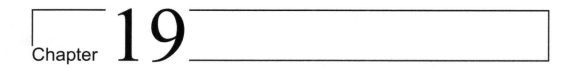

Chapter **19**

Antiderivatives of a Function

19.1 Questions about antiderivatives

- What is an antiderivative?

- Finding some antiderivatives.

- Functions with no elementary antiderivatives.

19.2 What is an antiderivative of a function?

19.2.1 Definition: ANTIDERIVATIVE

If g is a function and f is a function, on the same domain, such that $f'(x) = g(x)$ for all x in the domain, then f is one of the antiderivatives of g. That is, a function f is an antiderivative of g if f is a function whose derivative is g.

19.2.2 Example

If $g(x) = \cos(x)$, then one of the anti-derivatives of g is $\sin(x)$, because $\sin(x)$ is one of the functions whose derivative is $\cos(x)$. Another anti-derivative of $\cos(x)$ is $\sin(x) + 1$, and another is $\sin(x) + \frac{2}{\pi} + \frac{3}{2} + \frac{8-\pi}{5} + e^2$.

R. L. Moore, emphasizing the difference between the use of "the" and "a", stated it this way

There is no such thing as *the* function whose derivative is such and such, but *a* function whose derivative is such and such could exist.

19.2.3   Indefinite integral

Another name for an antiderivative is an *indefinite integral.* For example, an indefinite integral of x^2 is $x^3/3$ plus a constant. By abuse of notation, a question might be posed as follows: "What is the integral of x^2?" when the asker really means "what is an indefinite integral of x^2?" This violates the distinction between the indefinite article and the definite article and is objectionable for the reason stated in the R. L. Moore quote, above. We will stick to calling indefinite integrals, antiderivatives.

19.2.4   Problem

Prove that if a function $g(x)$ has an anti-derivative $f(x)$, then $g(x)$ has infinitely many anti-derivatives.

Solution

Suppose that a given function $g(x)$ has an anti-derivative $f(x)$, then

$$\frac{df(x)}{dx} = g(x).$$

But for every constant C, $f(x) + C$ is also an anti-derivative of $g(x)$, since

$$\frac{d}{dx}(f(x) + C) = \frac{d}{dx}f(x) + \frac{d}{dx}C = g(x) + 0 = g(x),$$

and this is true for infinitely many constants, C.

19.3   Hints for finding antiderivatives

It is not always easy to think of a function that is an antiderivative of a given function. Often, we look to see if there is something familiar enough in the expression for us to recognize that it is the derivative of something we know. A lot of time our success in finding an antiderivative is a hit or miss process involving trial and error, and checking the possible answer by differentiating it to see if it is indeed the function we have. However, there are some useful guideposts, which we state here as hints. Here they are.

1. Given a product of two functions, if you find that one of the factors is the derivative of something upon which the other function depends in a simple way, think of the chain rule.

2. Any number between 1 and -1 is the sine of some number.

3. If $b > 0$ and $b \neq 1$ then some power of b is a power of e.

4. Any number is the tangent of some number.

5. Any fraction is the sum of fractions.

19.3.1   Examples

Use the hints to find at least two anti-derivatives for each of the following functions

(a) $2x \cos(x^2)$

(b) $\frac{x+1}{(x-2)(x+3)}$

(c) $\cos(x)e^{\sin(x)}$

(d) $\cos^2(x) - \sin^2(x)$

(e) $\sec(x)$

(f) $-g'(x)\sin(g(x))$

(g) 7^x

Suggested solutions

(a) $2x$ is the derivative of x^2, and cos is the derivative of sin.

(b) The fraction $\frac{x+1}{(x-2)(x+3)}$ is $\frac{x+1}{x^2+x-6}$ and can be written as the sum of fractions,

$$
\begin{aligned}
\frac{x+1}{x^2+x-6} &= \frac{1}{2}\left(\frac{2x+1+1}{x^2+x-6}\right), \\
&= \frac{1}{2}\left(\frac{2x+1}{x^2+x-6} + \frac{1}{x^2+x-6}\right).
\end{aligned}
$$

These can further re-written to make the resulting fractions look like known derivatives. For example

$$\frac{1}{2}\left(\frac{2x+1}{x^2+x-6}\right) = \frac{1}{2}\left(\frac{\frac{d}{dx}\left(x^2+x-6\right)}{x^2+x-6}\right)$$

and

$$\frac{1}{2}\left(\frac{1}{x^2+x-6}\right) = \frac{1}{2}\frac{1}{(x-2)(x+3)},$$
$$= \frac{1}{10}\left(\frac{1}{x-2}\right) - \frac{1}{10}\left(\frac{1}{x+3}\right).$$

(c) Recall that $\cos(x)$ is the derivative of $\sin(x)$; think of the chain rule applied to the derivative of $e^{u(x)}$.

(d) Notice that $\cos^2(x) - \sin^2(x) = \cos(2x)$.

(e) Recall $\sec^2(x)$ and $\sec(x)\tan(x)$ are derivatives. Therefore their sum, $\sec^2(x)+\sec(x)\tan(x)$, is also a derivative. How is $\sec(x)$ related to $\sec^2(x)+\sec(x)\tan(x)$?

(f) $\cos(g(x))$, or $\cos(g(x)) + C$, where C is any constant.

(g) $7 = e^{\ln(7)}$.

19.3.2   Problem

If $g(x) = \cot(x)$, on the segment $(0, \pi)$, sketch four of its antiderivatives.

Solution

Since $y'(x) = \cot(x) = \frac{\cos(x)}{\sin(x)}$, then

$$y'(x) = \frac{\frac{d}{dx}\sin(x)}{\sin(x)} = \frac{d}{dx}\left[\ln\left(\sin(x)\right) + C\right]. \text{ So,}$$
$$y(x) = \ln\left(\sin(x)\right) + C.$$

Examples: Figure 19.1 shows a sketch for the four constants, $C = 0, \frac{3}{2}, 2, 3$.

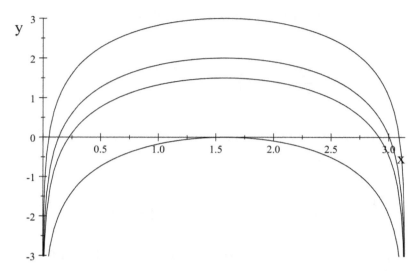

Figure 19.1 Graphs of $y = \ln\left(\sin(x)\right) + C$, for $C = 0, \frac{3}{2}, 2, 3$.

19.4 No elementary antiderivatives.

There are certain functions for which it is impossible to express an antiderivative in terms of "elementary functions". By *elementary functions* we mean a finite number of operations of addition, multiplication, subtraction, division, raising to a power, finding square roots of, taking logs, antilogs, trigonometric, anti-trigonometric or any finite compositions of such functions.

19.4.1 Example

Does there exist an antiderivative, expressed in terms of a finite number of elementary functions for

$$y'(x) = e^{x^2}?\tag{19.1}$$

Discussion

If we differentiate e^{x^2}, we get $2xe^{x^2}$. But Equation (19.1) does not have any multiple of x as a factor. If we experiment with various substitutions, we will see that we cannot find any finite combination of elementary functions satisfying this equation. Here is an example of a valiant attempt that fails. Let $x = r\sin(t)$, then $e^{x^2} = e^{r^2 \sin^2(t)}$. Rewrite the function $y(x)$ as a function $u(t)$. That is, $y(x) = y(r\sin(t)) = u(t)$; now set

$$\frac{dy}{dx} = \frac{du}{dt}\frac{dt}{dx} = e^{x^2}.$$

But what is $\frac{dt}{dx}$? From the equation $x = r\sin(t)$, we get

$$\frac{dt}{dx} = \frac{1}{r\cos(t)} = \frac{1}{\sqrt{r^2 - x^2}},$$

so

$$\frac{du}{dt}\frac{1}{r\cos(t)} = e^{r^2\sin^2(t)}.$$

Now if we could find a function $u(t)$ such that

$$u'(t) = r\cos(t)e^{r^2\sin^2(t)},$$

we would be done. Alas, this isn't possible because $u(t)$ is obviously a function of $e^{\sin^2(t)}$ and its derivative would be a multiple of $\sin(t)\cos(t)$, not just $\cos(t)$.

When we study functions which have non-elementary anti-derivatives we try to represent them in terms of infinite series or by the use of other special methods. For example, we can use the infinite series

$$e^x = 1 + x + \frac{x^2}{2!} + \frac{x^3}{3!} + ... + \frac{x^n}{n!} + ..., \qquad (19.2)$$

to write another infinite series for e^{x^2}.

$$e^{x^2} = 1 + x^2 + \frac{x^4}{2!} + \frac{x^6}{3!} + ... + \frac{x^{2n}}{n!} + ...$$

From this we can get an infinite sum of powers of x that is an antiderivative satisfying Equation (19.1).

19.5 Exercise 19 HOMEWORK

1. Find at least two antiderivatives of $\sqrt{x+7}$.

2. Find an antiderivative of
$$\frac{2x}{\sqrt{1-x^2}}.$$

3. Find a function $y(x)$ such that
$$\frac{d}{dx}y(x) = \cos^2(\sin(x))\cos(x).$$

4. Find an antiderivative of

   (a) $\frac{1}{x^2}e^{1/x}$.

   (b) $2xe^{x^2}$.

5. Find an antiderivative of $\cos^2(x)$.

6. Find an antiderivative of
$$\sqrt{1-x^2}.$$

7. Find an antiderivative of
$$x^2\sqrt{x^3-1}.$$

8. Find an antiderivative of

   (a) $\frac{2x^3}{\sqrt{1-x^4}}$.

   (b) $\frac{2x}{\sqrt{1-x^4}}$.

9. Find an antiderivative of

   (a) $xe^x + e^x$,

   (b) $x\sin(x)$. HINT: Find the derivative of $x\cos(x)$.

10. Find an antiderivative of
$$\csc(x).$$

11. Find an antiderivative of
$$\frac{1}{x^3 + 1}.$$

12. Find an antiderivative of the infinite series
$$1 + x^2 + \frac{x^4}{2!} + \frac{x^6}{3!} + \frac{x^8}{4!} + \ldots + \frac{x^{2n}}{n!} + \ldots.$$

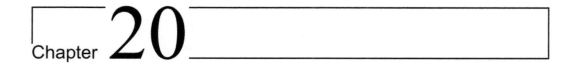

Chapter 20

Area Under a Curve

20.1 Questions about area

- What is a planar region?

- What is area?

- How is the area under the graph of a function related to an antiderivative of the function?

- What is an antiderivative of xe^x?

20.2 Planar regions

20.2.1 Definition CIRCULAR REGION

Let P be a point and r be a positive number and K be the circle with center at P and radius r. The statement that C *is the circular region with center at P and radius r* means that C is a point set such that a point X belongs to C if and only if the distance from X to P is less than r.

In other words, C is the interior of the circle K. In Figure 20.1 we show an example of a circle K and the circular region C.

20.2.2   Definition SQUARE REGION

Let the points A, B, C, and D be the vertices of a square. The square region S is a point set such that a point X belongs to it if and only if X belongs to some segment with one endpoint on one side of the square and the other end on the opposite side. In other words a square region is the interior of a square. In Figure 20.1, we show an example of a square $ABCD$ and the square region S.

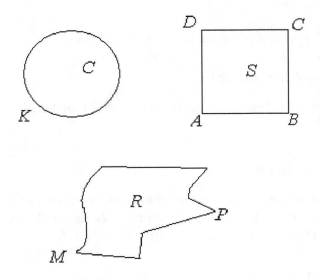

Figure 20.1 Planar Regions.

20.2.3   A closed curve M and a region R

Intuitively, we can roughly describe a *simple closed curve* as a collection of straight line segments and curved arcs joined together in such as way as to enclose some set of points in the plane. This is done by having the arcs start at some point P, then have the connected pieces return to that same point P as the ending point. In Figure 20.1 we exhibit an example of a simple closed curve M. The point set that is the interior of M is the region R.

20.3 Area

We will take the word *area* to be an undefined term. But any concept of it must satisfy the following requirements:

1. Area is a non-negative number assigned to a region.

2. The area of the union of two non-over-lapping regions is additive.

3. If a figure is translated from one position in the plane to another without distorting the figure the area is unchanged.

4. The area of a unit square is 1.

5. The area of a single point or of a segment is zero.

Example: Area of a rectangle.

Given a rectangle with base of length b and height of length h, the area is the number $b \times h$.

20.3.1 Subregions

Since a region is a set, we may use the definition of subset to define a subregion. The statement that A *is a subregion of* B means both A and B are regions and that there is no point in A that is not in B.

20.3.2 Problem: Area of a subregion

Show that if A is a subregion of the region B, then the area of A is less or equal to than the area of B.

Solution

Let A and B be sets with $A \subseteq B$. Let $B - A$ be the set such that $x \in B - A$ if and only if x is in B, but not in A. The interior of $B - A$ is a region, and the interior of A is a region and these two regions do not overlap. Let $m(A)$ be the area of A and $m(B - A)$ be the area of $B - A$. The region $A \cup (B - A)$ is B, and the area is additive. Thus, $m(A) + m(B - A) = m(B)$.

But both $m(A)$ and $m(B-A)$ are nonnegative, therefore, $m(B) \geq m(A)$ and $m(B) \geq m(B-A)$.

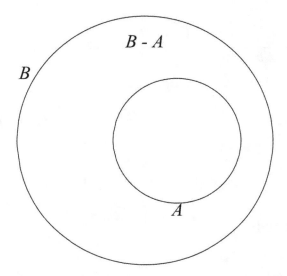

Figure 20.2 Area of B greater than Area of A.

20.3.3   Example of subregions formed by simple graphs

We will be studying areas of regions formed by a closed curve consisting of a given function, an interval on the x-axis and vertical intervals, as shown in Figure 20.3.   Region 2 is the region bound by the graph F, the x-axis, and vertical line segments at $x = a$, and $x = b$. Region 1 is a rectangle that is a subset of Region 2, and Region 3 is a rectangle for which Region 2 is a subset. The areas for these three regions are indicated. The area of Region 1 is less than the area of Region 2 which is less than the area of Region 3.

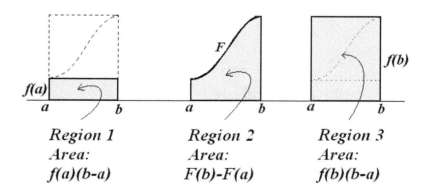

Figure 20.3 Comparison of Areas.

20.3.4 Example

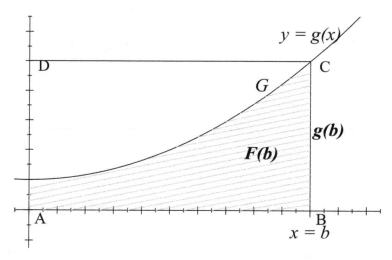

Figure 20.4. The area under $g(x) = x^2 + 1$ from 0 to b.

Let us look at a region with area $F(b)$ formed by the graph, G, of the function $g(x) = x^2 + 1$, the *x-axis*, the *y-axis*, and the vertical line $x = b$, for any $b > 0$. See Figure 20.4. This region is a subset of the rectangle $ABCD$, whose area is $b \times g(b)$, hence the area of the given region is not greater than the area of the rectangle.   In other words,

$$F(b) \leq b \times g(b). \tag{20.1}$$

Consider the same graph and let a be a positive number less than b. The area under the curve and between the y-axis and $x = a$ is $F(a)$. What is the area of the region under G above the x-axis and between the two vertical lines $x = a$ and $x = b$?   Answer: $F(b) - F(a)$.

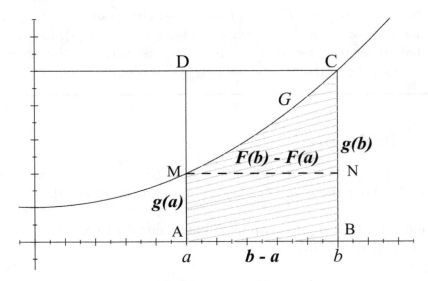

Figure 20.5 The area $F(b) - F(a)$.

The area $F(b) - F(a)$ is less than the area of the rectangle $ABCD$, and greater than the area of the rectangle $ABNM$.   That is,

$$g(a) \times (b - a) \leq F(b) - F(a) \leq g(b) \times (b - a), \text{ for all } a < b. \tag{20.2}$$

The region shown in Figure 20.5 will always be between the two rectangles and if we divide the inequality (20.2) by $(b - a)$, we will get

$$g(a) \leq \frac{F(b) - F(a)}{b - a} \leq g(b) \tag{20.3}$$

for all $a < b$.

20.3.5   Problem

What happens to (20.3) as $b \to a$?

Solution

In the Inequality (20.3), the expression

$$\frac{F(b) - F(a)}{b - a}$$

approaches the derivative $F'(a)$ as $b \to a$, and the quantity $g(b)$ approaches $g(a)$, while $g(a)$ stays $g(a)$. Therefore, we get $F'(a) = g(a)$, for any a; this means that the area under the curve is an anti-derivative of the curve. Thus for all $x > 0$ in the domain of G, we may write $F'(x) = g(x)$.

20.3.6   Problems

1. Using $g(x) = x^2 + 1$, and the fact that $F'(a) = g(a)$, find the exact area $F(2.5)$ under the graph G between 0 and a in Figure 20.4.

2. Find the area, approximately to one decimal place, under G between $x = 1.3$ and $x = 2.7$.

Solutions

Since $F'(a) = g(a) = a^2 + 1$, then $F(a) = a^3/3 + a + C$. But notice that $F(0) = 0$; therefore, $C = 0$. This means that $F(a)$ is

$$F(a) = a^3/3 + a. \tag{20.4}$$

We may now answer the questions 1 and 2.

1. From Equation (20.4), we can say that $F(2.5) = \frac{185}{24}$.

2. The area between 1.3 and 2.7 is $F(2.7) - F(1.3) \approx 9.3 - 2.0 = 7.3$.

20.4   Areas and antiderivatives

20.4.1   Problem

Let $y = g(x)$ be the equation of a positive, non-increasing, and continuous graph G on an interval $[a, b]$.   Now let $x \in [a, b]$ and let $F(x)$ be the area between a and x.   Give an argument that

$$\frac{dF(x)}{dx} = g(x). \tag{20.5}$$

.

Solution

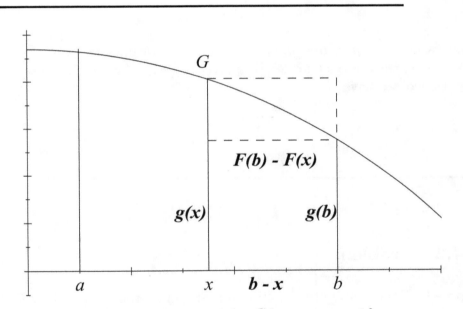

Figure 20.6 Region under G between a and b.

If $F(x)$ is the area under G, above the x-axis, and between a and x, and $F(b)$ is the area between between a and b, then $F(b) - F(x)$ is the area between x and b for all numbers, b.   Notice that, since $g(x)$ is non-increasing, the area $F(b) - F(x)$ is always greater than or equal to the area $g(b)(b - x)$ and less than or equal to the area $g(x)(b - x)$, that is

$$g(x)(b - x) \geq F(b) - F(x) \geq g(b)(b - x). \tag{20.6}$$

This can be seen by drawing two rectangles both with base $b - x$ and one with height $g(x)$ and the other with height $g(b)$. Now, with $b \neq x$, divide both sides by $b - x$, getting

$$g(x) \geq \frac{F(b) - F(x)}{b - x} \geq g(b). \qquad (20.7)$$

As the number b approaches x the above inequalities (20.7) remain true and $g(b) \to g(x)$. But the quotient

$$\frac{F(b) - F(x)}{b - x} \qquad (20.8)$$

is always between $g(b)$ and $g(x)$, so it approaches $g(x)$ as well. However, when $b \to x$, the quotient in (20.8) approaches $F'(x)$ by definition of derivative. Hence, we have

$$F'(x) = g(x).$$

20.4.2   Problem

Suppose G is the graph of a function $y = \sin(x)$ on the interval $(0, \pi)$. Find the area of the region under G, above the x-axis, and between the vertical lines $x = 0$ and $x = \pi$.

Solution

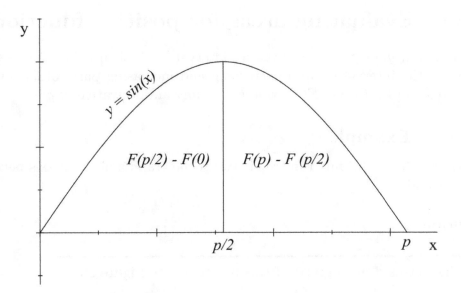

Figure 20.7 Area under the sine curve, from 0 to π. Here, $p = \pi$.

From $F'(x) = \sin(x)$, we get $F(x) = -\cos(x) + C$. Now since $F(0) = 0$, we can write $0 = -\cos(0) + C$. Therefore $C = 1$. This means

$$F(x) = 1 - \cos(x).\tag{20.9}$$

Using Equation (20.9), we have

$$
\begin{aligned}
F(\frac{\pi}{2}) - F(0) &= 1 - \cos(\frac{\pi}{2}) = 1, \\
F(\pi) - F(\frac{\pi}{2}) &= (1 - \cos(\pi)) - (1 - \cos(\frac{\pi}{2})) = 1.
\end{aligned}
$$

Hence the area under one loop of the sine curve is 2.

We could have found the area equally well by using the antiderivative $F(x) = -\cos(x) + C$ over the entire interval $[0, \pi]$ without finding a value for the constant C and without breaking the graph up into increasing and decreasing parts. That is, the area is

$$
\begin{aligned}
F(\pi) - F(0) &= -\cos(\pi) + C - (-\cos(0) + C) \\
&= -(-1) + C + 1 - C = 2.
\end{aligned}
$$

20.5   Evaluating areas for positive functions

If a function $g(x) \geq 0$ for all x in an interval $[a, b]$, then it is not necessary to break the interval up into decreasing and increasing parts of the graph. The area will be $F(b) - F(a)$, where F is any antiderivative of $g(x)$.

20.5.1   Example

If $g(x) = -3x^4 + 3x^2$ find the area between the graph and the x-axis between $x = -1$ and $x = 1$.

Solution

First let us draw a graph of this function. See Figure 20.8

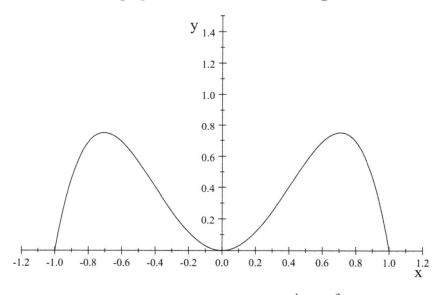

Figure 20.8 Graph of $y = -3x^4 + 3x^2$.

Let $F(x)$ be an antiderivative of $-3x^4 + 3x^2$, then

$$F(x) = \frac{-3}{5}x^5 + x^3 + C.$$

The area is $F(1) - F(-1)$ which is

$$\frac{-3}{5} + 1 + C - \left(\frac{-3}{5}(-1) + (-1) + C \right) = \frac{4}{5}.$$

If you compute the two areas under the graph, over the intervals $[-1, 0]$ and $[0, 1]$, you will see that each one is $\frac{2}{5}$.

20.6   An antiderivative of xe^x

Suppose we wanted to find the area of the region between $x = 0$ and $x = 2$, above the x-axis and under the graph whose equation is $y = xe^x$? See Figure 20.9.

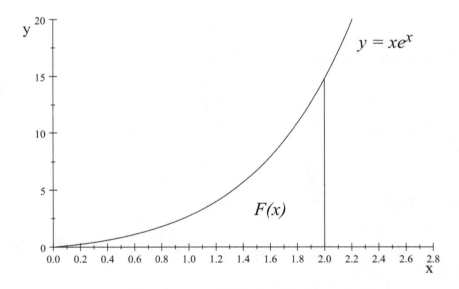

Figure 20.9 Area under the graph of $y = xe^x$.

We know that the derivative of the area $F(x)$ is given by

$$F'(x) = xe^x. \tag{20.10}$$

What is an antiderivative of xe^x?   At this point we introduce an old mathematician's saying

"A mathematician doesn't learn anything new by being logical all the time". R. L. Moore December 1955.

I believe that this puzzling statement was meant to encourage us to try something counter-intuitive. Here for example, we wanted the antiderivative of xe^x. We notice that its derivative produces xe^x, itself as one of its terms. So maybe we can use differentiation to find an antiderivative. That is, try $(xe^x)' = xe^x + e^x$. This motivates the following discussion.

In Equation (20.10), xe^x is a function x times the derivative of another function e^x. That is,

$$xe^x = x\frac{de^x}{dx}.$$

Now if we could reverse this and find a function, say $M(x)$, whose derivative is the second function times the derivative of the first, namely

$$M'(x) = e^x\frac{dx}{dx},$$

we could add these equations and get

$$xe^x + M'(x) = x\frac{de^x}{dx} + e^x\frac{dx}{dx}. \tag{20.11}$$

The right hand side of Equation (20.11) is the derivative of the product xe^x.

$$\frac{d}{dx}(xe^x) = x\frac{de^x}{dx} + e^x\frac{dx}{dx}. \tag{20.12}$$

Then by equations (20.11) and (20.12), we have

$$xe^x = \frac{d}{dx}(xe^x) - M'(x), \text{ so} \tag{20.13}$$

$$F'(x) = \frac{d}{dx}(xe^x) - M'(x), \text{ or} \tag{20.14}$$

$$F'(x) = \frac{d(xe^x)}{dx} - e^x. \tag{20.15}$$

That is,

$$F(x) = xe^x - e^x + C,$$

which can be easily checked by differentiating $F(x)$.

The constant can be determined by noticing that the area under the graph at $x = 0$ is itself 0. That is

$$F(0) = 0,$$
$$C = 1.$$

The area under the curve from 0 to 2 is

$$
\begin{aligned}
F(2) - F(0) &= 2e^2 - e^2 + 1 - (0e^0 - e^0 + 1) \\
&= e^2 + 1, \\
&\approx 8.389.
\end{aligned}
$$

20.7   Integration by parts formula

Sometimes the process of finding an antiderivative of a function is called *finding an integral of* the function.   (We will discuss integrals in a later chapter.)   In other words, if $y'(x) = g(x)$, then an antiderivative $y(x)$ is called an integral of $g(x)$.   Finding the integral of a function is also called integration of that function.

One method for finding an antiderivative of a function such as xe^x is called *integration by parts,* which is based upon the formula for the derivative of a product.

If u and v are two differentiable functions on the same domain then $(uv)' = uv' + vu'$.   An integration by parts formula could then be $uv' = (uv)' - vu'$.

Professor Moore said it this way

> If you have a function expressed as one function times the derivative of another function, find a function whose derivative is the other times the derivative of the one.   (April 1956)

20.7.1   Problem

Use integration by parts to find an antiderivative of $x \cos(x)$.

Solution

What we have is $x \cos(x)$.   We note that $\cos(x)$ is the derivative of $\sin(x)$.   That is, $x \cos(x) = x \frac{d}{dx} \sin(x)$.   We want to find two functions u and v such that

$$
\frac{d}{dx}(uv) = u\frac{d}{dx}v + v\frac{d}{dx}u.
$$

Let's try $v = \sin(x)$ and $u = x$, then

$$\frac{d}{dx}(x\sin(x)) = x\cos(x) + \sin(x).$$

Solving for $x\cos(x)$, we get

$$\begin{aligned}
x\cos(x) &= \frac{d}{dx}x\sin(x) - \sin(x), \\
&= \frac{d}{dx}x\sin(x) + \frac{d}{dx}\cos(x).
\end{aligned}$$

This means that $x\sin(x) + \cos(x)$ is an antiderivative of $x\cos(x)$.

20.8 Another old mathematician's useful trick

"Any number between 1 and -1 is the sine of some angle" -R. L. Moore. (March, 1956)

20.8.1 Problem

Find an antiderivative $y(x)$ when

$$\frac{dy(x)}{dx} = \sqrt{1 - x^2}.$$

Solution

Clearly x is between 1 and -1, therefore it must be the sine of some angle t. Let $x(t) = \sin(t)$, then

$$y(x) = y(\sin(t)), \tag{20.16}$$

$$\begin{aligned}
\sqrt{1 - x^2} &= \cos(t), \text{ and} \\
t &= \sin^{-1}(x).
\end{aligned}$$

Equation (20.16) indicates that $y(x)$ can be written as a function of t. Let's use $w(t)$ for $y(\sin(t))$, that is

$$w(t) = y(x). \tag{20.17}$$

Differentiate $w(t)$ in Equation (20.17) with respect to t, using the chain rule and we get

$$\frac{dw}{dt} = \frac{dy}{dx}\frac{dx}{dt}, \text{ or}$$

$$\frac{dw}{dt} = \sqrt{1 - x^2}\cos(t),$$

$$\frac{dw}{dt} = \cos^2(t). \qquad (20.18)$$

But

$$\cos^2(t) = \frac{1}{2}(\cos(2t) + 1), \text{ so}$$

$$\frac{dw}{dt} = \frac{1}{2}\cos(2t) + \frac{1}{2}.$$

Hence, an antiderivative is

$$w(t) = \frac{1}{4}\sin(2t) + \frac{1}{2}t + C,$$

$$w(t) = \frac{1}{4}2\sin(t)\cos(t) + \frac{1}{2}t + C.$$

Therefore, after replacing t by $\sin^{-1}(x)$ and $w(t)$ by $y(x)$,

$$y(x) = \frac{1}{2}\sin(\sin^{-1}(x))\cos(\sin^{-1}(x)) + \frac{1}{2}\sin^{-1}(x) + C,$$

or

$$y(x) = \frac{1}{2}x\sqrt{1 - x^2} + \frac{1}{2}\sin^{-1}(x) + C.$$

20.9   Exercise 20 HOMEWORK

1. Sketch the graph of $y = 1 - \cos(x)$ between $x = 0$ and $x = \pi$ and find the area below the graph and above the x-axis on the interval $[0, \pi]$.

2. Sketch the graph of $y = \sin(x) + 1$ and find the area under this graph above the x-axis and between the vertical lines $x = 0$ and $x = 2\pi$.

3. Let $h(x) = 1 + x^2 + x^4/2$ be the equation of a graph H defined on the interval $[0, \frac{3}{2}]$. Find the area under H, above the x-axis and between $x = 0$ and $x = \frac{3}{2}$.

4. Find an antiderivative of xe^{2x}.

5. If $F(x)$ is a function such that $F'(x) = x \sin(x)$, find $F(x)$.

6. Let $r > 0$ and G be a quarter-circle whose equation is

$$y = \sqrt{r^2 - x^2} \text{ for } 0 \le x \le r.$$

   (a) If H is the plane region under G, above the x-axis and between $x = 0$ and $x = r$, find the area of H.

   (b) Use this to show that the area of a circle with radius r is πr^2. Sketch a graph illustrating this solution.

7. Let a and b be positive numbers and let E be the ellipse whose equation is

$$\frac{x^2}{a^2} + \frac{y^2}{b^2} = 1,$$

find the area enclosed by E.

8. Let $y = g(x)$ be the equation of a positive continuous graph G on an interval $[a, b]$. See Figure 20.10.

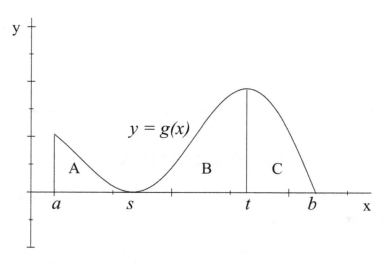

Figure 20.10 Area under an oscillating graph.

Suppose G is increasing on part of the interval and decreasing on part of the interval. Find the areas of regions A, B, and C in terms of an antiderivative of $g(x)$. What is the area under the entire graph from a to b?

9. If the graph of $y = g(x)$ is below the x-axis, then we need to find the antiderivative of $-y$, in order to get a positive number for the area. Let G be the graph of the equation $y = x^2 - 1$, on the interval $[-2, 2]$. Find the area between the x-axis and G. Use $-y$ to get the area of the parts below the x-axis. See Figure 20.11. Treat all regions $A, B, C,$ and D as having positive areas.

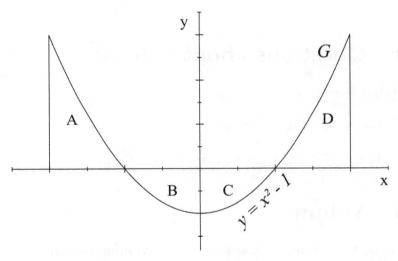

Figure 20.11 Use $-y$ to get positive areas for regions B and C.

Chapter 21

Volume

21.1 Questions about volume

- What is volume?

- Volumes of solids of revolution.

- Volumes of solids with known cross-sections.

21.2 Volume

A solid, or three-dimensional region, has volume measured in cubic units. Volume is taken as an undefined term that obeys the following properties:

1. Volume is a non-negative number assigned to a solid region.

2. The volume of the union of two non-over-lapping regions is additive.

3. If a solid figure is translated from one position in three-dimensional space to another without distorting the figure the volume is unchanged.

4. The volume of a unit cube is 1.

5. The volume of a single point, a segment or a two-dimensional region is zero.

21.3   Solids of revolution

In three-dimensional space, if a simple graph G is rotated about the x-axis, we get a solid whose cross sections are circles, obtained by taking "slices" perpendicular to the x-axis.

21.3.1   Example: The solid obtained by rotating the graph of $y = x^2$.

Let G be the graph defined by $y(x) = x^2$, and for any number $p > 0$, define the plane region $R(p)$ as the point set between $x = 0$ and $x = p$, below G and above the x-axis. See Figure 21.1.

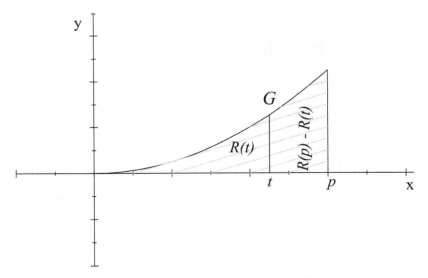

Figure 21.1 The plane region $R(p)$.

If we rotate $R(p)$ about the x-axis we get the solid $S(p)$ shown in Figure 21.2. The cross sections of $S(p)$ are circles.

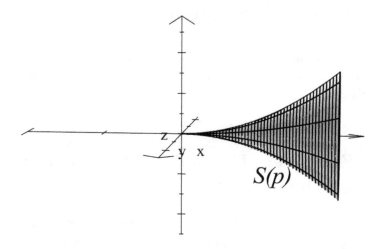

Figure 21.2. Solid of revolution $S(p)$.

21.3.2 Question: What is the area of a cross section?

For any number t such that $0 < t < p$ what is the area of the cross section of $S(p)$ at $x = t$?

Answer

For any number t, the graph G has an ordinate $y(t) = t^2$. Therefore the circular cross section at $x = t$ will have a radius of t^2. The area of that cross section will be $\pi \times (t^2)^2$ or πt^4.

21.3.3 Example: Volume of $S(p)$ between $x = 0$ and $x = p$.

See Figure 21.3.

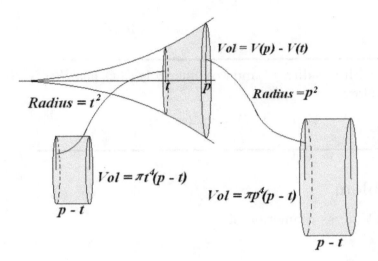

Figure 21.3 The volume between two cylinders.

21.3.4 Problem

If $0 < t < p$, show that

$$\pi t^4 (p - t) \leq V(p) - V(t). \tag{21.1}$$

Solution

Since $V(p) - V(t)$ is the volume of the solid $S(p) - S(t)$ and πt^4 is the area of the circular cross section at $x = t$, then there is a cylinder with a circular base of πt^4 that is a subset of the solid $S(p) - S(t)$.

21.3.5 Problem

The volume $V(p) - V(t)$ is less than $\pi p^4 (p - t)$, why?

Answer

The cylinder with a radius p^2 and volume $\pi p^4(p - t)$ has $S(p) - S(t)$ as a subregion, therefore.

$$V(p) - V(t) \leq \pi p^4(p - t). \tag{21.2}$$

21.3.6 Problem

Find the volume $V(x)$ as a function of x.

Answer

From inequalities (21.1) and (21.2), we get

$$\pi t^4(p - t) \leq V(p) - V(t) \leq \pi p^4(p - t). \tag{21.3}$$

We can divide by $p - t$, since it is positive,

$$\pi t^4 \leq \frac{V(p) - V(t)}{p - t} \leq \pi p^4,$$

and this is true for all $0 < t < p$. If we let $p \to t$, then

$$\lim_{p \to t} \frac{V(p) - V(t)}{p - t} \to \pi t^4.$$

In other words,

$$\frac{dV(t)}{dt} = \pi t^4. \tag{21.4}$$

Equation (21.4) tells us that the volume $V(t)$ is an antiderivative of πt^4. So,

$$V(t) = \frac{1}{5}\pi t^5 + C, \text{ for all } t.$$

When $t = 0$, the volume is 0. Thus $V(0) = 0 \Rightarrow C = 0$.
The volume at any x is

$$V(x) = \frac{1}{5}\pi x^5.$$

21.4 Solids with known cross sections

We have just seen that $S(p)$ is a solid whose cross sections are circles with varying radii.   We found the volume $V(x)$ at any x by noting that the volume $V(x) - V(t)$ was between the volumes of two circular cylinders, one with radius $y(t)$ and one with radius $y(x)$.

$$\pi y^2(t)(x - t) \le V(x) - V(t) \le \pi y^2(x)(x - t), \qquad (21.5)$$

which leads us to the conclusion that $\frac{dV}{dx} = \pi y^2(x)$. In words,

> The volume of a solid of revolution of a function $f(x)$ about the x-axis is an antiderivative of $\pi \left(f(x) \right)^2$.

Now we want to work with solids which when "sliced" at some point along their axis will have cross sections that are not necessarily circles, but whose areas can be determined.

21.4.1 Example

In Figure 21.4, we show a diagram of a dam whose cross sections are triangles. The height of each triangle is a constant and the base is a line segment in the xy-plane.   The base is perpendicular to the y-axis and its length is determined by a curve in the xy-plane.

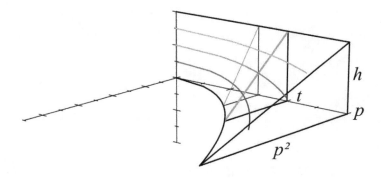

Figure 21.4 Dam with trianglular cross sections.

These are right-angle triangles. For any given number y there is a triangle whose right angle is the point $(0, y, 0)$. The fixed number h, is the height of each triangle. Each triangle has its third point on the curve $y^2 = x$ in the first quadrant of the xy-plane. The triangle whose right angle is at $(0, y, 0)$ has an area of

$$\frac{1}{2}y^2 h.$$

The volume from $y = 0$ to $y = p$ is $V(p)$. The volume of the solid between $y = p$ and $y = t$ is shown as $V(p) - V(t)$ in Figure 21.5

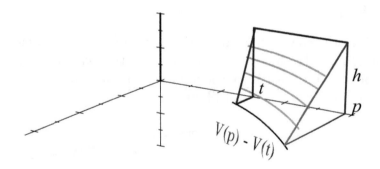

Figure 21.5 Volume of the dam between $y = t$ and $y = p$.

This volume, $V(p) - V(t)$, will be less than the volume, $\frac{1}{2}p^2 h(p - 1)$, of the parallelepiped whose face is the triangle at $y = p$ shown in Figure 21.6

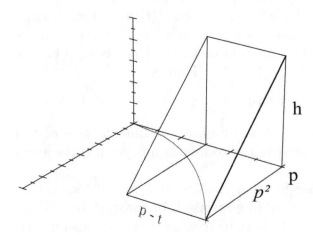

Figure 21.6 Parallelepide with face p^2.

In other words,

$$V(p) - V(t) \leq \frac{1}{2}p^2 h(p - t).$$

Constructing a similar parallelepiped whose face is the smaller triangle at $y = t$ gives us a volume of $\frac{1}{2}t^2 h(p - t)$. This means that the following inequality is true for all p and t when $0 < t < p$.

$$\frac{1}{2}t^2 h(p - t) \leq V(p) - V(t) \leq \frac{1}{2}p^2 h(p - t). \tag{21.6}$$

Dividing Inequality (21.6) by $(p - t)$ yields

$$\frac{1}{2}t^2 h \leq \frac{V(p) - V(t)}{p - t} \leq \frac{1}{2}p^2 h. \tag{21.7}$$

The limit as $p \to t$ gives us $V'(t) = \frac{1}{2}t^2 h$, so the volume out to t is an antiderivative of $\frac{1}{2}t^2 h$,

$$V(t) = \frac{1}{6}t^3 h + C.$$

Here, since $V(0) = 0$, then $C = 0$.

In general, let S be a three-dimensional figure with length p. Let the length be measured by a variable x that runs from 0 to p. If a is any positive number $0 \le a \le p$, let $A(a)$ be the area of the cross sectional slice at $x = a$. If $A(t) < A(a)$ for all $t < a$, then $A(t)(a - t) < A(a)(a - t)$. Let $V(t)$ be the volume of S, for any $t > 0$, then $V(a) - V(t)$ is the volume of S between $x = a$ and $x = t$. This volume $V(a) - V(t)$ itself, is between the two volumes $A(t)(a - t)$, and $A(a)(a - t)$. That is,

$$A(t)(a - t) \le V(a) - V(t) \le A(a)(a - t). \tag{21.8}$$

Dividing by $a - t$ and finding the limit as $t \to a$, results in $\frac{dV}{da} = A(a)$. Thus, the derivative of the volume at any point is the area of the cross section at that point.

On the other hand, if $A(t) > A(a)$ for all $t < a$, then the inequality in Equation (21.8) would be reversed, but we still have $V'(a) = A(a)$. In other cases the solid might have increasing cross sections for some part and decreasing cross sections for other parts, again the volume at a is an antiderivative of the cross section at a.

21.5 Exercise 21 HOMEWORK

1. Let $r > 0$ and G be the graph of the equation

$$y = \sqrt{r^2 - x^2}.$$

Let H be the planar region under G above the x-axis and between $x = 0$ and $x = r$. Let S be the solid obtained by rotating H about the x-axis. Find the volume of S, and use this to show that the volume of a sphere of radius r is $\frac{4}{3}\pi r^3$. Illustrate this solution with a sketch.

2. Let a and b be positive numbers, and let G be the ellipse defined by the equation

$$\frac{x^2}{a^2} + \frac{y^2}{b^2} = 1.$$

Find the volume of the ellipsoid obtained by rotating this ellipse about the x-axis.

3. If G is the ellipse defined in Problem 2, find the volume of the ellipsoid rotated about the y-axis.

4. Suppose a tree has a cylindrical trunk of radius r. If a wedge is cut at a 30° angle find the volume of the wedge. See Figure 21.7.

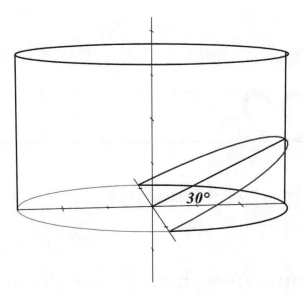

Figure 21.7. Volume of a wedge.

5. Find the volume of the intersection of two right circular cylinders at right angles to each other. Sketch a graph.

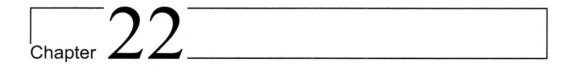

Chapter 22

Density and Pressure

22.1 Some questions about density

- The weight of an object with uniform density.

- The weight of an object with varying density.

- Hydrostatic pressure.

22.2 Uniform density

The density of an object is defined, in general, as its mass per unit volume. In a gravitational field such as the Earth's gravity, we can also define density as the force acting on the mass (the weight) per unit volume. We will confine our study to the weight-density.

By "uniform density",- we mean that the object has the same ratio of weight/volume over every possible subset of the object.

22.2.1 Problem

Suppose a marble table top has a uniform density of d pounds per cubic foot. If the slab has two straight edges and one edge is a curve defined by $y = \sqrt{x}$ as shown in Figure 22.1, find the weight in terms of d if the length x is 1 foot and the thickness z is one inch. If the density of marble is about 170 lbs. per cubic foot, find the weight of the table top in pounds.

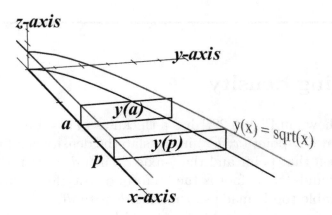

Figure 22.1 $y(x) = \sqrt{x}$.

Solution

Since $z = 1/12$, the cross sectional rectangles at any x are $\sqrt{x} \times 1/12$. The volume at x is $V(x)$, and the weight is $W(x) = V(x) \times d$. We call this the "weight-slab" because it has the volume being multiplied by the density. The weight between cross sections at $x = a$ and $x = p$, where $p > a$, is $W(p) - W(a)$, which satisfies the inequality

$$\sqrt{a} \times 1(p-a)d/12 \le W(p) - W(a) \le \sqrt{p} \times 1(p-a)d/12. \qquad (22.1)$$

Dividing by $(p - a)$ and finding the limit as $p \to a$, we get $W'(a) = d\sqrt{a}/12$, or, $W'(x) = dx^{1/2}/12$ for all x,

$$W(x) = \frac{d}{18}x^{3/2} + C.$$

Since $W(0) = 0$, then $C = 0$. The density was given as d pounds per cubic foot and the length is 1 foot, so

$$W(1) = \frac{2d}{3}(1)^{3/2} \times \frac{1}{12} = \frac{d}{18} \text{ cubic feet.}$$

Using the density of 170 pounds per cubic foot, this works out to be about 9.4 lbs.

22.3 Varying density

Suppose the slab shown in Figure 22.1 has a thickness of $z = 1$ and a varying density, such that at any point (x, y, z) in the slab the density is a function of the distance between that point and the y-axis. Then $d = d(x)$, not just d. For example if d pounds/cubic foot is the uniform density for a cubic foot of material, but the table top is made so that the density $d(x) = f(x)d$ where $f(x)$ is a function of x, then the table top would have a varying density.

22.3.1 Problem

If the slab in Figure 22.1, with $z = 1$, has a density at each point (x, y, z) that is equal to one-half of the distance, x from the (y, z)-plane. Let d pounds per cubic foot be the density of a uniform block of the material. Find the weight of the slab.

Solution

See Figure 22.2. For this figure the volume for any x is $V(x)$, so between $x = a$ and $x = p$, the volume is $V(p) - V(a)$ and satisfies the following inequality

$$y(a) \times (p - a) \le V(p) - V(a) \le y(p) \times (p - a).$$

To find the weight of the slab, we multiply the density $d(x) = 0.5x \times d$ times the area of the cross section to get $y(x) \times 0.5x \times d$. This is equivalent to making the slab have a cross section of $\sqrt{x} \times 0.5x \times d$. So now the weight-slab looks like it has an increasing height of $0.5x \times d$, rather than just a height of 1.

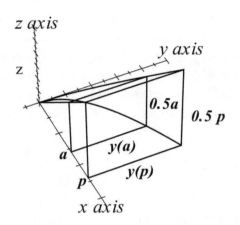

Figure 22.2 The weight slab has cross-section $\sqrt{x} \times 0.5x \times d$.

See Figure 22.2. The volume times density at x is the weight, $W(x)$. So the value $W(p) - W(a)$ is less than $\sqrt{p} \times d(p) \times (p - a)$ and greater than $\sqrt{a} \times d(a) \times (p - a)$ for all $p > a$. Here $d(x) = 0.5xd$, so

$$\sqrt{a} \times 0.5a \times d \times (p - a) \;\leq\; W(p) - W(a) \leq \sqrt{p} \times 0.5p \times d \times (p - a),$$

$$0.5a \times \sqrt{a} \times d \;\leq\; \frac{W(p) - W(a)}{p - a} \leq 0.5p \times \sqrt{p} \times d. \qquad (22.2)$$

As $p \to a$, inequality (22.2) tells us that

$$\frac{dW(a)}{da} = 0.5a^{3/2}d,$$

or for any x

$$\frac{dW(x)}{dx} = 0.5x^{3/2}d.$$

So,

$$W(x) = \frac{1}{5}x^{5/2}d + C,$$

and since $W(0) = 0$, then $C = 0$. The weight when $x = 1$ is

$$W(1) = \frac{1}{5}d \text{ pounds.}$$

In this problem we were given that the density varied with x, and was $0.5x \times d$ pound per cubic foot. If the material is marble with a density of $d = 170$ pounds per cubic foot, the weight will be $\frac{1}{5} \times 170 = 34$ lbs.

What this example shows is that, although an object may have a constant height, and has a varying density, we can solve the problem as if it had a constant density and the height varied. The density of a three-dimensional solid may vary in one, two, or all three of its dimensions.

22.3.2   Example

Suppose that a given rectangular table top has its length measured along the x-axis and width measured along the y-axis and thickness or height along the z-axis. The table top is made of material that would have a uniform density of d pounds per cubic foot.

Now suppose this table top is made so that the density varies along the length and across the width of the table according to the equation $d(x, y) = x^2 y^3$. This means that if x is the measurement, in feet along the length of the table, and y is the measurement across the width, then density at the point (x, y, z) is $x^2 y^3 d$ pounds per cubic foot. We summarize this by saying that the three-dimensional table top has a density of $x^2 y^3 d$ at each of its points, (x, y, z).

22.3.3   Problem

If a rectangular table top is 5 feet by 3 feet, by 1 foot thick and is made so that the density at any point is $x^2 y^3$ times d pound per cubic foot, where x is the measurement in feet along the length and y is the measurement in feet across the width, find the weight of the table top. Let us also assume the density does not vary in the vertical direction. Find the weight of this table top.

Solution

See Figure 22.3.

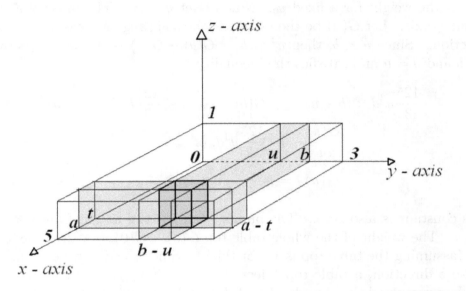

Figure 22.3 Table top with point density $x^2 y^3 d$.

Let t and a be two numbers on the x-axis, with $0 \leq t < a \leq 5$. Let u and b be two numbers on the y-axis with $0 \leq u < b \leq 3$. For any fixed y_0, let $F(x)$ be the weight of the solid region of the table top when $x \in$ the interval $[0, 5]$ on the x-axis. Now $F(a) - F(t)$ is the weight of the solid region between $x = t$ and $x = a$. Since $t < a$, then $t^2 < a^2$, so the following inequality holds

$$t^2 y_0^3 d \times (a - t) \leq F(a) - F(t) \leq a^2 y_0^3 d \times (a - t).$$

Dividing by $(a - t)$ and finding the limit as $a \to t$, we get

$$\frac{dF(t)}{dt} = t^2 y_0^3 d, \text{ or}$$

$$F(t) = \frac{t^3}{3} y_0^3 d + C.$$

The constant $C = 0$ because $F(0) = 0$.

$$F(5) - F(0) = \frac{125 y_0^3 d}{3}.$$

This is the weight for a fixed y_0. Now we let y vary. The interval $[0, u]$ is on the y-axis. Let $G(u)$ be the weight of the solid region measured in the y direction. Since $u < b$, then $u^3 < b^3$. $G(b) - G(u)$ is the weight between $y = b$ and $y = u$ and satisfies the inequality

$$\frac{125}{3}u^3 d \times (b - u) \ \leq \ G(b) - G(u) \leq \frac{125}{3}b^3 d \times (b - u)$$

$$\frac{dG(u)}{du} \ = \ \frac{125}{3}u^3 d, \text{ or}$$

$$G(u) \ = \ \frac{125}{3}\frac{u^4}{4}d + C.$$

This constant is also zero. The number $G(u)$ is the weight from $y = 0$ to $y = u$. The weight of the whole table top $G(3) - G(0) = \frac{125}{3}\frac{81}{4}d = 843.75d$ lbs. (assuming the table top is 1 foot thick). Since the density did not vary in the z direction, a table top z feet thick it weighs $843.75d \times z$ lbs. This problem presumed that the density of d pound per cubic foot varied as $x^2 y^3$. If the material were marble, the answer would have to be multiplied by 170 pounds.

22.4 Hydrostatic pressure

In physics, pressure is defined as Force /Area. In our discussion we take the weight of an object to be the force caused by the Earth's gravity. So, when we define pressure, P, caused by the weight, W, of an object that sits on a planar region of area A, we can write the equation

$$P = \frac{W}{A}.$$

We always assume that the force (weight) is being applied perpendicular to the region.

22.4.1 Example

If a bucket contains two gallons of water, what is the pressure on the bottom of the bucket? Answer: W/A, where W is the weight of two gallons of water and A is the area of the bottom of the bucket. Unfortunately, this "mathematicians answer" is not so much an answer as it is a way to tell you how to get the answer, if you only had the data required to complete the solution.

22.4.2   Problem

If water has a density of 62.2 pounds per cubic foot and the bottom of a bucket is a circle with an 8 inch diameter and the water in the bucket is deep enough to hold 2 gallons, what is the pressure on the bottom in terms of pounds per square inch?   Assume the bucket has vertical walls; that is, it is not tapered.

Solution

First, how many gallons are in a cubic foot? Then, what is the area of the bottom in square feet? The easy part is the area; it is $16\pi\ in^2 \approx 50.27.in^2$.

We know that $144\ in^2 = 1\ ft^2$. The bottom is approximately $A = 0.35$ $sq.\ ft$. The number of gallons in one cubic foot is approximately 7.48. Since we were given that one cubic foot of water weighs 62.2 pounds, then 2 gallons weigh $W = \frac{2}{7.48} \times 62.2 \approx 16.6$ pounds. This means that the force on the bottom of the bucket is 16.6 pounds and the pressure $P = \frac{16.6}{0.35} \approx 47.43$ pounds per square foot.

Sometimes the volume is not given in terms of gallons, but the depth of the water is given, so you will be required to calculate the volume in gallons. Also you may be interested in the pressure on the side of a container.

22.4.3   Example

Suppose you are given a tank with flat sides, (not a round bucket) with water in it. If, at a depth of h feet from the surface of the water, there is some region, R_1, with area A is on the bottom of the tank, we can find the force (weight) on R_1 by knowing the volume of the water above the region. This is because the weight is the density times the volume. After finding the weight, we can use this to find the pressure (weight per unit area). See "Tank 1" in Figure 22.4. If the the area A is $(1/2\ ft)(1/2\ ft)$, and the depth h, is 9 inches ($\frac{3}{4}\ ft$), the volume is $\frac{3}{16}ft^3$. The density of water being 62.2 lbs per

ft^3 makes the weight 11.6635 *lbs* of water applied to a 1/4 ft^2 region, so the pressure is 46.65 pounds per square foot

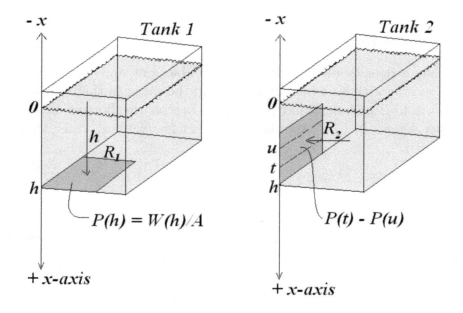

Figure 22.4 Pressure on bottom and side of a tank.

Suppose the 1/2 *ft* by 1/2 *ft* region is on the side of the tank and the water is 3/4 *ft* deep, what do you think the pressure will be? To solve this note that the water depth varies from the top of the region to the bottom.

22.4.4   Problem

Find the force (weight) on R_2, a 1/2 by 1/2 *ft* square region on the *side* of the tank with the bottom edge of the square on the bottom of the tank, the top edge on the side of the tank 3 inches below the surface of the water. After finding the weight find the pressure in pounds per square foot. See "Tank 2", Figure 22.4.

Solution

Let t and u be two numbers on the *downward pointing* x-axis, with $0 \leq u < t \leq h$. Here h is $\frac{3}{4}$ of one foot. The area of the small subregion of R_2 (between $x = u$ and $x = t$) is $(t - u)\, ft \times \frac{1}{2}\, ft$. The top of R_2 is $1/4$ ft. from the water surface, so for $x \geq 1/4$, the weight at depth x is $W(x)$. The weight between u and t is $W(t) - W(u)$. But this weight on the small subregion is less than the weight, $W(t)$, at depth of t and it is greater than the weight, $W(u)$, at a depth of u, so

$$\frac{1}{2}u(t - u)62.2 \leq W(t) - W(u) \leq \frac{1}{2}t(t - u)62.2.$$

Divide by $(t - u)$, getting

$$31.1u \leq \frac{W(t) - W(u)}{t - u} \leq 31.1t.$$

Taking the limit as $u \to t$, yields

$$\frac{dW(t)}{dt} = 31.1t.$$

This makes

$$W(t) = 31.1\frac{t^2}{2} + C = 15.5t^2 + C$$

We want the weight on the entire region R, which is

$$W(\frac{3}{4}) - W(\frac{1}{4}) = 15.5\left((\frac{3}{4})^2 - (\frac{1}{4})^2\right) = 7.75 \text{ lbs.}$$

The pressure is $\frac{W}{A}$, or

$$\frac{7.75\ lbs}{0.25\ ft^2} = 31 \text{ lbs./ft.}^2$$

22.5   Exercise 22 HOMEWORK

1. A 5 foot by 3 foot rectangular table top with a thickness of z feet is made so that it has a point density of $d(x, y, z) = x^2 d$ lbs. per cubic foot, where $0 \leq x \leq 5$ is the measurement in feet on the long side of the table top. Find the weight of the table top.

2. Let S be a slab (a table top) that is five feet long in the x direction and three feet wide in the y direction, and 4 inches in the z direction. If the density of S at each of its points (x, y, z) is $d(x, y) = xy^2d$, find the weight of the table top.

3. If the density of a 5 foot by 3 foot by 1/2 foot table top has a point density of $d(x, y, z) = x^2y^3z^4d$ lbs. per cubic foot, find the weight. The length is measured along the x-axis, width along the y-axis and thickness along the z-axis.

4. A water tank with planar sides has a region, R, on one of the sides. The region R is bound by the downward pointing x-axis from $x = 0$ to $x = 1$ and the graph of $y = \sqrt{x}$. The water in the tank is one foot deep. In Figure 22.5, R is the darker shaded region and $t - u$ is the length of the interval $[u, t]$.

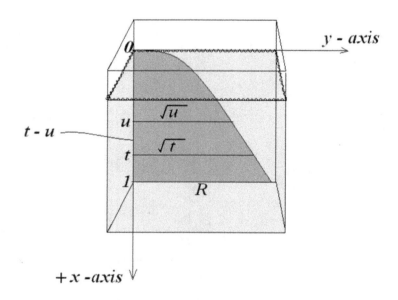

Figure 22.5 Pressure on the parabolic region, R (shaded).

(a) Find the force (weight in pounds) on R.

(b) Find the pressure in pounds per square foot on the region.

5. Given a circular gate of radius 2 feet on the side of a planar tank of water with the center of the circle 4 feet below the surface of the water, find the force (weight) on the upper half of the gate and on the lower half. Find the pressure in pounds/square foot on each half.

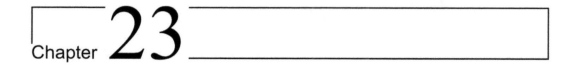

Chapter 23

Arc Length

23.1 Questions about arc length

- What is an antiderivative of $1/(1-x^2)^2$?

- What is an antiderivative of $\sec^3(x)$?

- What is an antiderivative of $\sqrt{1+x^2}$?

- The length of an arc vs. length of a chord.

- Arc length formula.

- Elliptic Integrals.

A difficult problem encountered in ancient Greek mathematics was that of finding the length of a curve. Some examples would be the problems of finding the perimeter of a circle, or the length of one loop of the sine wave, or the length of an ellipse, or of a segment of a parabola. Fortunately, the invention of calculus went a long way toward solving some of these problems, although not all of them. At least one curve in this list still cannot be solved by ordinary calculus tools.

But even for those that we can solve, the procedure is often difficult and convoluted. If you were just given the solutions, you would not appreciate how difficult these problems are. Here, you will enjoy the satisfaction of deriving, for yourselves, the solutions "from scratch". This means that you will use only your basic knowledge of algebra, trigonometry and the calculus you have already learned.

First, let us consider three preliminary problems, whose solutions are needed in order to find arc lengths of certain curves. Find an antiderivative of

1. $1/(1 - x^2)^2$.

2. $\sec^3(\theta)$.

3. $\sqrt{1 + x^2}$.

23.2   What is an antiderivative of $1/(1 - x^2)^2$?

23.2.1   Problem

Find a function $y(x)$ such that

$$\frac{dy(x)}{dx} = \frac{1}{(1 - x^2)^2}. \tag{23.1}$$

Solution

Rewrite the fraction in Equation (23.1) as

$$\frac{1}{(1 - x)^2(1 + x)^2} \tag{23.2}$$

We will recall a method from algebra called *resolving a fraction into partial fractions*, which, in this case, requires us to find constants a, b, c, and d, so that the following expression is an identity true for all x.

$$\frac{1}{(1 - x)^2(1 + x)^2} = \frac{ax + b}{(1 - x)^2} + \frac{cx + d}{(1 + x)^2}. \tag{23.3}$$

Equation (23.3) can be written as

$$1 = (ax + b)(1 + x)^2 + (cx + d)(1 - x)^2.$$

Then by letting x be (in turn) the four different values $x = 1, x = -1, x = 0$, and $x = 2$, we get the following four equations.

$$
\begin{aligned}
4(a + b) &= 1 \\
4(-c + d) &= 1, \\
b + d &= 1, \\
18a + 9b + 2c + d &= 1.
\end{aligned}
$$

Solving this system, we get: $a = -1/4$, $b = 1/2$, $c = 1/4$, $d = 1/2$. Therefore we may write Equation (23.3) as

$$
\frac{1}{(1 - x)^2(1 + x)^2} = \frac{(-1/4)x + (1/2)}{(1 - x)^2} + \frac{(1/4)x + 1/2}{(1 + x)^2},
$$

which we will re-write as

$$
\begin{aligned}
\frac{1}{(1 - x)^2(1 + x)^2} &= \frac{1}{4}\left(\frac{-x + 2}{(1 - x)^2} + \frac{x + 2}{(1 + x)^2}\right) \\
&= \frac{1}{4}\left(\frac{1 - x + 1}{(1 - x)^2} + \frac{1 + x + 1}{(1 + x)^2}\right), \\
\frac{dy(x)}{dx} &= \frac{1}{4}\left(\frac{1}{1 - x} + \frac{1}{(1 - x)^2} + \frac{1}{1 + x} + \frac{1}{(1 + x)^2}\right) \text{ (23.4)}
\end{aligned}
$$

An antiderivative $y(x)$ is

$$
y(x) = \frac{1}{4}\left(-\ln(1 - x) + \frac{1}{1 - x} + \ln(1 + x) - \frac{1}{1 + x}\right) + C,
$$

which can be re-written as

$$
y(x) = \frac{x}{2(1 - x^2)} + \ln\left(\frac{1 + x}{1 - x}\right)^{1/4} + C. \tag{23.5}
$$

This is the solution to Equation (23.1).

23.3 What is an antiderivative of $\sec^3(\theta)$?

23.3.1 Problem

Find a function $g(\theta)$, such that

$$
\frac{dg(\theta)}{d\theta} = \sec^3(\theta).
$$

Solution

Before looking any further into this solution, try re-writing $\sec(\theta)$ in terms of $\cos(\theta)$.

Write $\sec^3(\theta)$ as $1/\cos^3(\theta)$.

$$\sec^3(\theta) = \frac{1}{\cos^3(\theta)},$$

$$\sec^3(\theta) = \frac{1}{\cos^4(\theta)} \cos(\theta).$$

Since cosine is the derivative of the sine, let us re-write this as a function of the sine times the derivative of the sine,

$$\sec^3(\theta) = \frac{\cos(\theta)}{(1 - \sin^2(\theta))^2}.$$

If we use $\cos(\theta)$ for x in Expression(23.2), then $\cos(\theta)/(1 - \sin^2(\theta))^2$ can be written as

$$\frac{\cos(\theta)}{4} \left(\frac{1}{1 - \sin(\theta)} + \frac{1}{(1 - \sin(\theta))^2} + \frac{1}{(1 + \sin(\theta))} + \frac{1}{(1 + \sin(\theta))^2} \right).$$

Therefore, by Equations (23.4) and (23.5),

$$g(\theta) = \frac{\sin(\theta)}{2(1 - \sin(\theta)^2)} + \ln\left(\frac{1 + \sin(\theta)}{1 - \sin(\theta)} \right)^{1/4} + C. \qquad (23.6)$$

We can check that this is a function whose derivative is $\sec^3(\theta)$.

23.4   What is an antiderivative of $\sqrt{1 + x^2}$?

Unlike the expression $\sqrt{1 - x^2}$ which has a restriction on x, namely that x must neither be greater than 1, nor less than -1, here x can be any number. Recall the Old Mathematician's saying that "...any number is the tangent of some angle," we will assume that $x = \tan(\theta)$, for some angle θ.

23.4.1 Problem

Find a function $y(x)$ such that

$$\frac{dy(x)}{dx} = \sqrt{1+x^2}. \tag{23.7}$$

Solution

Let $x = \tan(\theta)$, and sketch a right triangle with an angle θ whose tangent is x. This also yields $\sqrt{1+x^2} = \sec(\theta)$. See Figure 23.1.

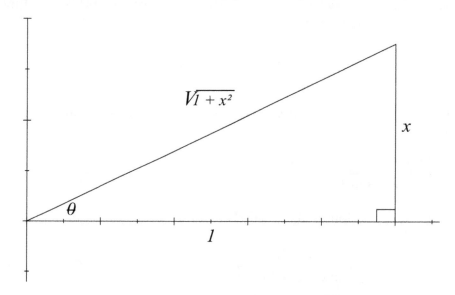

Figure 23.1 Triangle for $x = \tan(\theta)$.

When we write $y(x) = y(\tan(\theta))$ we are defining a function of θ, so we let $y(\tan(\theta)) = u(\theta)$. Differentiating with respect to x, we get

$$\frac{d}{dx}y(x) = \frac{d}{d\theta}u(\theta)\frac{d\theta}{dx}. \tag{23.8}$$

Now differentiate the equation $x = \tan(\theta)$ with respect to x, getting $1 = \sec^2(\theta)\frac{d\theta}{dx}$, or

$$\frac{d\theta}{dx} = \frac{1}{\sec^2(\theta)}. \tag{23.9}$$

And since $\sqrt{1+x^2} = \sec(\theta)$, Equations (23.7)-(23.9) tell us that

$$\frac{d}{d\theta}u(\theta)\frac{1}{\sec^2(\theta)} = \sec(\theta), \text{ or}$$

$$\frac{d}{d\theta}u(\theta) = \sec^3(\theta). \qquad (23.10)$$

From Equation (23.6), our solution to the $\sec^3$ problem, we have

$$u(\theta) = \frac{\sin(\theta)}{2(1-\sin(\theta)^2)} + \ln\left(\frac{1+\sin(\theta)}{1-\sin(\theta)}\right)^{1/4} + C.$$

But now we've got to change this back to an expression in terms of x, rather than θ. Using Figure 23.1, we get

$$y(x) = \frac{1}{2}x\sqrt{1+x^2} + \ln\left(\frac{\sqrt{1+x^2}+x}{\sqrt{1+x^2}-x}\right)^{1/4} + C.$$

When we check this we will see that the derivative of $y(x)$ is $\sqrt{1+x^2}$.

23.5   Length of an arc versus length of a chord

Given a differentiable function $f(x)$, and two points A and B on the graph of $y = f(x)$, we want to find the distance between A and B as measured along the graph itself (the length of the arc $\overset{\frown}{AB}$) and the length of the straight line segment (the chord $\overline{AB}$).

23.5.1   Example

Let f be the function $f(x) = \frac{1}{2}x^2$, and F be the graph of the equation

$$y = f(x).$$

On the graph F, let A and X be two points whose abscissas are a and x, respectively. The length of the piece of this parabola between A and X is called the *arc* $\overset{\frown}{AX}$ and is measured along the curve from A to X, as shown in Figure 23.2.

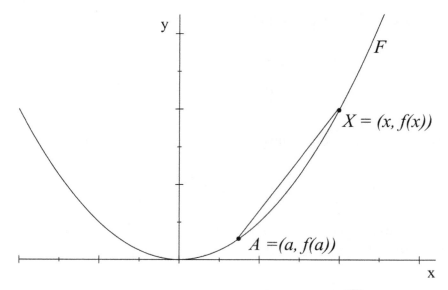

Figure 23.2 The chord $\overline{AX}$, and the arc $\overset{\frown}{AX}$.

We want to find the length of the arc $\overset{\frown}{AX}$. To solve this problem we will make an assumption (stated later) about the ratio $\overset{\frown}{AX}/\overline{AX}$ of the arc length to the chord length, and what happens to this ratio as $A \to X$.

23.5.2   Problem

If F is the graph of the equation

$$y = \frac{1}{2}x^2,$$

and A and X are two points of F with abscissas a and x, respectively, find the length of the chord $\overline{AX}$ and the arc $\overset{\frown}{AX}$.

By the Pythagorean Theorem, the chord $\overline{AX}$ has length

$$\overline{AX} = \sqrt{(x-a)^2 + (f(x) - f(a))^2}.$$

Let us denote the arc length of F as follows. Trace the curve F from the origin to a point $X = (x, f(x))$, and let $S(x)$ be the length measured along this curve. The length of the arc $\overset{\frown}{AX}$ will be $S(x) - S(a)$. That is,

$$\overset{\frown}{AX} = S(x) - S(a). \tag{23.11}$$

Let $\overline{AX}$ be the chord joining points A and X and write the ratio $\widehat{AX}/\overline{AX}$ as

$$\frac{\widehat{AX}}{\overline{AX}} = \frac{S(x) - S(a)}{\sqrt{(x-a)^2 + (f(x) - f(a))^2}}$$

$$\frac{\widehat{AX}}{\overline{AX}} = \frac{S(x) - S(a)}{(x-a)\sqrt{1 + \left(\frac{f(x) - f(a)}{x-a}\right)^2}}. \tag{23.12}$$

Equation (23.12) can be re-written as

$$\frac{S(x) - S(a)}{x - a} = \frac{\widehat{AX}}{\overline{AX}}\sqrt{1 + \left(\frac{f(x) - f(a)}{x - a}\right)^2}. \tag{23.13}$$

We will assume without proof that the ratio of the arc length to the chord length approaches 1 as $A \to X$. That is,

$$\lim_{A \to X} \frac{\widehat{AX}}{\overline{AX}} = 1.$$

As the point A approaches the point X, then the abscissa $a \to x$, thus

$$\lim_{a \to x} \frac{S(x) - S(a)}{x - a} = \lim_{A \to X} \frac{\widehat{AX}}{\overline{AX}} \lim_{a \to x} \sqrt{1 + \left(\frac{f(x) - f(a)}{x - a}\right)^2}. \tag{23.14}$$

What does Equation (23.14) become?

$$\frac{dS(x)}{dx} = \sqrt{1 + \left(\frac{df(x)}{dx}\right)^2}. \tag{23.15}$$

Here, the function is $f(x) = \frac{1}{2}x^2$, so, $f'(x) = x$, and Equation (23.15) is

$$\frac{dS(x)}{dx} = \sqrt{1 + x^2}, \text{ or}$$

$$S(x) = \frac{1}{2}x\sqrt{1 + x^2} + \ln\left(\frac{\sqrt{1 + x^2} + x}{\sqrt{1 + x^2} - x}\right)^{1/4} + C.$$

The length of the arc from A to B is $S(x) - S(a)$.

23.5.3   Problem

In Figure 23.2, if the coordinates of A and X are $(1, \frac{1}{2})$ and $(2, 2)$, find the arc length $\widehat{AX}$, and the chord length $\overline{AX}$.

Solution

The arc length is

$$
\begin{aligned}
\widehat{AX} &= S(2) - S(1) \\
&\approx 2.95787 - 1.14779, \\
&\approx 1.81012.
\end{aligned}
$$

The chord length is

$$
\begin{aligned}
\overline{AX} &= \sqrt{1 + (3/2)^2} \\
&\approx 1.80278.
\end{aligned}
$$

23.6   Arc length formula

In general, if $f(x)$ is *any* function with a continuous derivative on an interval $[a, p]$, and $A = (a, f(a))$, $P = (p, f(p))$, then the arc length $\widehat{AP}$ is $S(p) - S(a)$, where $S(x)$ is an antiderivative of $\sqrt{1 + \left(\frac{df(x)}{dx}\right)^2}$. This is the same formula given in Equation (23.15) as derived in the special case, above This formula works even in cases where the graph of $f(x)$ both curves up and curves down on the given interval.

23.6.1   Problem

If $f(x)$ is a function with a continuous derivative and having a relative maximum and a relative minimum on the interval $[a, b]$, as shown in Figure 23.3, derive the formula for the arc length $\widehat{AB}$.

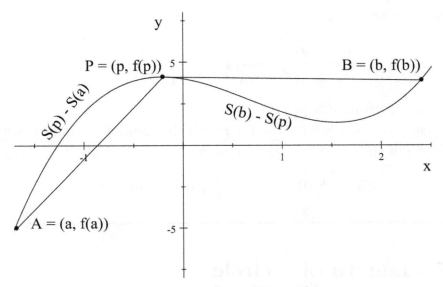

Figure 23.3 $\widehat{AB} = \widehat{AP} + \widehat{PB}$.

Solution

We consider the two lengths $\widehat{AP}$ and $\widehat{PB}$.

$$
\begin{aligned}
\widehat{AP} &= S(p) - S(a) \text{ and} \\
\widehat{PB} &= S(b) - S(p).
\end{aligned}
$$

Find the length of the chord $\overline{AP}$.

$$\overline{AP} = \sqrt{(p-a)^2 + (f(p) - f(a))^2}.$$

The ratio of the arc $\widehat{AP}$ to the chord $\overline{AP}$ is

$$\frac{\widehat{AP}}{\overline{AP}} = \frac{S(p) - S(a)}{\sqrt{(p-a)^2 + (f(p) - f(a))^2}}.$$

Factor $(p-a)$ out from the radical and rewrite this equation as

$$\frac{S(p) - S(a)}{p - a} = \frac{\widehat{AP}}{\overline{AP}} \sqrt{1 + \left(\frac{f(p) - f(a)}{p - a}\right)^2}.$$

The limit as $A \to P$ is

$$\frac{dS(p)}{dp} = \sqrt{1 + \left(\frac{df(p)}{dp}\right)^2}. \tag{23.16}$$

Find an antiderivative $S(p) + C$.

Repeat this process for the arc $\widehat{PB}$ and the chord $\overline{PB}$, finding $S(b) + C$. The length of the arc $\widehat{AB}$ is $\widehat{AP} + \widehat{PB}$, or

$$S(b) - S(a) = (S(b) - S(p)) + (S(p) - S(a)).$$

23.7 Length of a circle

We will derive the formula for the circumference of a circle.

23.7.1 Problem

Let C be a circle with center at the origin and radius, r, a positive number. Find the circumference of the circle using Equation (23.15).

Solution

The equation of C is $x^2 + y^2 = r^2$. Let us write y as a function of x on the interval $[-r, r]$

$$y(x) = \pm\sqrt{r^2 - x^2}.$$

Consider only the top half of the circle. If $S(x)$ is the arc length, and since $y'(x) = -x/\sqrt{r^2 - x^2}$, then by Equation (23.15),

$$\begin{aligned}
\frac{dS(x)}{dx} &= \sqrt{1 + (y'(x))^2} \\
&= \sqrt{1 + \frac{x^2}{(r^2 - x^2)}}, \\
&= \frac{r}{\sqrt{r^2 - x^2}},
\end{aligned}$$

an antiderivative of which is

$$S(x) = r \sin^{-1}(\frac{x}{r}) + C. \qquad (23.17)$$

Therefore the arc length on $[-r, r]$, (one-half of the circle) is $S(r) - S(-r) = r\sin^{-1}(1) - r\sin^{-1}(-1) = \pi r$. Thus, the whole circle has circumference $2\pi r$.

23.8   Length of an ellipse.

An ellipse has length, just as a circle has a perimeter, but as we shall see, we encounter significant difficulties when we try to use Equation (23.15) to find the length of a (non-circular) ellipse.

23.8.1   Problem

Let a and b be two positive numbers with $a \neq b$, and let E be the graph of the equation

$$\frac{x^2}{a^2} + \frac{y^2}{b^2} = 1.$$

Find the length of the ellipse.

Solution

Solve for y as a function of x,

$$y(x) = \pm\frac{b}{a}\sqrt{a^2 - x^2}.$$

Consider only the top half of the ellipse for $x \in [-a, a]$. If $S(x)$ is the arc length and since $y'(x) = -bx/(a\sqrt{a^2 - x^2})$, then

$$\frac{dS(x)}{dx} = \sqrt{1 + (y'(x))^2}$$

$$= \sqrt{1 + \frac{b^2 x^2}{a^2(a^2 - x^2)}},$$

$$\frac{dS(x)}{dx} = \frac{\sqrt{1 - (a^2 - b^2)x^2/a^4}}{\sqrt{1 - x^2/a^2}}. \qquad (23.18)$$

If $a = b$, then Equation (23.18) reduces to the problem of finding the length of a circle, because both the axes of the ellipse are the same, becoming the radius r. If $a > 0$, and $b = 0$, then the equation reduces to $\frac{dS(x)}{dx} = 1$, or $S(x) = x + c$. The "degenerate" ellipse (one of the axes $=$ zero) has length $S(a) - S(-a) = 2a$. In all other cases, Equation (23.18) presents us with a dilemma. We can try to use a sine or cosine substitution in one of two ways: first, let's try

$$x = \frac{a^2}{\sqrt{a^2 - b^2}} \sin(\theta). \qquad (23.19)$$

Then

$$\frac{dS(x)}{dx} = \frac{\sqrt{1 - \sin^2(\theta)}}{\sqrt{1 - (a^2/(a^2 - b^2)) \sin^2(\theta)}}. \qquad (23.20)$$

Let $k = a/\sqrt{a^2 - b^2})$, then Equation (23.20) becomes

$$\frac{dS(x)}{dx} = \frac{\sqrt{1 - \sin^2(\theta)}}{\sqrt{1 - k^2 \sin^2(\theta)}}. \qquad (23.21)$$

Equation (23.21) cannot be solved by a finite combination of additions, subtractions, multiplications or divisions of the elementary functions. It's solution can be obtained only by numerical approximations obtained from a function called an *elliptic integral*.

Let us try another sine substitution in Equation (23.18). Suppose

$$x = a \sin(\theta),$$

then Equation (23.18) becomes

$$\frac{dS(x)}{dx} = \frac{\sqrt{1 - h^2 \sin^2(\theta)}}{\sqrt{1 - \sin^2(\theta)}}, \qquad (23.22)$$

where $h = \left(\sqrt{a^2 - b^2} \right)/a$. So Equation (23.22) is also an *elliptic integral*.

The value of an elliptic integral can be obtained by one of many different "numerical" methods, which we will not cover in this book. The term *Numerical Method* refers to some approximating method that uses infinite series, or some other formula in which the derivative is replaced by a quotient of "finite differences," and employs repeated steps, leading to approximate values. Often such approximations are fed back into the procedure to refine the accuracy of the approximation. The numerical process is usually programmed into a computer for rapid calculation.

23.9   Exercise 23 HOMEWORK

1. The hyperbolic sine (the $\sinh(t)$) and the hyperbolic cosine (the $\cosh(t)$) are functions based upon the graph of a hyperbola. These functions satisfy the following equations.

$$\sinh(t) = \frac{e^t - e^{-t}}{2}$$

$$\cosh(t) = \frac{e^t + e^{-t}}{2}.$$

   (a) Find the derivative of $\sinh(t)$, and the derivative of $\cosh(t)$.

   (b) Show that $\cosh^2(t) = 1 + \sinh^2(t)$, for all t.

   (c) If $A = (-3, \cosh(-3))$ and $B = (3, \cosh(3))$, find the length of the arc $\widehat{AB}$.

2. If G is the graph of $y(x) = 2x^{2/3}$, find the length of an arc on G between $x = 1$ and $x = 8$.

3. Find the length of the graph of $y(x) = \ln(\cos(x))$, from $x = \pi/6$ to $x = \pi/4$.

4. Assume that it is not possible to find an antiderivative, in elementary functions, of $\sqrt{1 - k^2 \sin^2(\theta)}$, when $k \neq 1$. Show that it is not possible to find the length of one loop of the sine curve from 0 to π in elementary functions.

5. Find an antiderivative of $\sec^2(2\theta)$. That is, find a function $y(\theta)$ such that $y'(\theta) = \sec^2(2\theta)$.

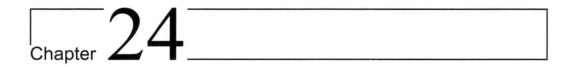

Chapter **24**

Integral Notation

24.1 Questions about integrals

- Geometric quantities.

- Approximating sums.

- The definite integral.

- The mean value theorem for integrals.

- The derivative of the integral.

24.2 Geometric quantities

In the last four chapters of this book we have been studying geometric quantities such as

1. Area, $A(x)$.

2. Volume, $V(x)$.

3. Density, $d(x)$.

4. Weight, $W(x)$.

5. Pressure, $P(x)$.

6. Length, $S(x)$.

In each of these cases we assumed that for the graph F of some function $f(x)$, the quantity could be measured in terms of x on an interval $[0, x]$. Take area, $A(x)$ for example, where we saw that if $t > x$, and we measured the area $A(t)$ under the graph of $y = f(x)$ from 0 to t, then we found that the area between x and t, $A(t) - A(x)$, satisfies the inequality,

$$f(x)(t - x) \le A(t) - A(x) \le f(t)(t - x), \qquad (24.1)$$

if F was increasing on $[x, t]$. Or, if F was decreasing on $[x, t]$, then

$$f(x)(t - x) \ge A(t) - A(x) \ge f(t)(t - x). \qquad (24.2)$$

Thus, either

$$f(x) \le \frac{A(t) - A(x)}{t - x} \le f(t)$$

or

$$f(x) \ge \frac{A(t) - A(x)}{t - x} \ge f(t).$$

In either case, we see that the limit as $t \to x$, yields

$$\frac{dA(x)}{dx} = f(x).$$

From this we are able to find the area $A(x)$ by finding an antiderivative of $f(x)$.

Another method for finding area, volume, density etc., is to compute an approximation of the quantity being sought, then to carry out a limiting process that yields that particular quantity. We start by observing that area, volume, density, etc. are obtained as *products* of numbers. For example the area is the product of two one-dimensional measures, width, w, times height, h; that is, $A = w \times h$. Volume is area times length, $V = A \times l$. Pressure is force divided by area (the product of force and the reciprocal of area), and as we shall see, *work* is force times distance.

If these problems involve only products or quotients, why do we need the calculus to solve them? The answer is that the quantities being multiplied together are not *constants*. For example in the table top problems we saw that density varied at points according to their distances from the edges of the table top. Another example is that the volume of a solid of revolution over an interval $[a, b]$, varies according to the areas of the "slices" (the circular cross sections) as we move along the interval from a to b.

An approximating method that can be used for solving such a problem is to divide $[a, b]$ into small sub-intervals and assume that the cross sectional area of the solid is a constant over each small subinterval, making the solid over that given subinterval a small cylinder. We will then compute the volume of these small cylinders, and add them up. The resulting answer will be an approximation of the volume. This "approximating sum" can then be improved by subdividing the subintervals themselves into smaller subintervals, *refining the partition of* $[a, b]$, getting more and smaller cylinders, thus reducing the error of the approximation.

For a given function $f(x)$ on an interval $[a, b]$, the method of *approximating sums* gives us an advantage in that it allows us to use well-known summation formulas and properties. The approximating sums approach a *number* called the "integral" of $f(x)$ on $[a, b]$.

On the other hand, the advantage of using the method described in Inequalities (24.1) and (24.2) is that we can immediately see that the quantity is an *antiderivative* of $f(x)$ on the interval $[a, b]$.

It turns out that both methods yield the same results.

24.3 The definite integral $\int_a^b f(x)dx$

24.3.1 DEFINITION: Partition of an interval $[a, b]$.

Let $[a, b]$ be an interval on the x-axis and n be a positive integer. Let $\mathcal{P}_n[a, b]$ be a collection of n non-overlapping subintervals filling up $[a, b]$. By non-overlapping intervals, we mean that their interiors don't overlap. Adjacent intervals can have an endpoint in common. The right endpoint of one interval is the same point as the left endpoint of the next interval.

To be unambiguous, we will use "uniform" partitions. That is, each subinterval in the collection will have the same length. When the partition is uniform, we can find the length of each subinterval simply by dividing the length of the original interval $b - a$, by the number n of subintervals. Thus each subinterval has length $\frac{b-a}{n}$.

24.3.2   EXAMPLE

Let $[a, b]$ be the interval $[3, 8]$ and let $n = 10$, find the uniform partition $\mathcal{P}_{10}[3, 8]$.

Solution

The interval $[3, 8]$ has length 5 and since we want to subdivide it into $n = 10$ subinterval, each one will be of length $\frac{5}{10}$; therefore the collection of such subintervals is

$$P_{10}[3, 8] = \{[3.0, 3.5], \ [3.5, 4.0], \ [4.0, 4.5], \ ..., \ [7.5, 8.0]\}.$$

We label the endpoints of each subinterval by subscripts, i, running from $i = 0$, to $i = 10$. Thus $x_0 = 3.0$, $x_1 = 3.5$, $x_2 = 4.0$, ..., $x_9 = 7.5$, $x_{10} = 8.0$. See Figure 24.1.

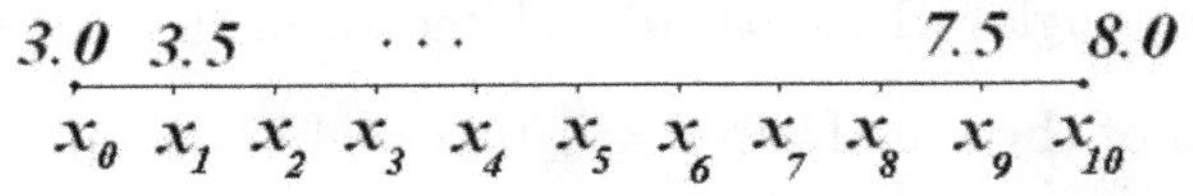

Figure 24.1 The uniform partition $\mathcal{P}_{10}$ of $[3, 8]$.

24.3.3   Problem

Let F be the graph of the equation $y = \sqrt{x}$ on the interval $[1, 4]$. Let $n = 6$ and $\mathcal{P}_6[1, 4]$ be the uniform partition $\{[1, 1.5], [1.5, 2.0], ..., [3.5, 4.0]\}$.

Find an *approximation* of the area under F, above the x-axis and between $x = 1$ and $x = 4$ by using the six rectangles constructed as follows. The bases of the rectangles are the subintervals in the partition, $\mathcal{P}_6[1, 4]$, and the heights are the ordinates of points on F computed at the abscissas of the left endpoints of the subinterval.

Solution

Let's draw a graph and construct rectangles whose bases are the intervals in the partition and whose heights are the values of $f(x)$ computed at the left end points of the intervals in the partition. See Figure 24.2

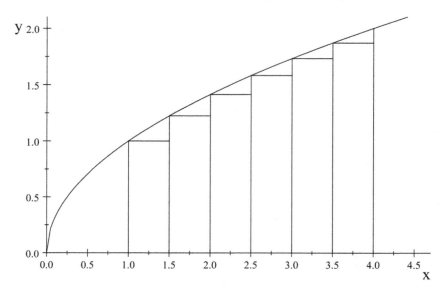

Figure 24.2 Rectangles whose bases are intervals of the partition.

An approximation of the area from $x = 1$ to $x = 4$ is the sum of the areas of the rectangles, that is

$$A \approx \sum_{i=1}^{6} \sqrt{x_{i-1}}(x_i - x_{i-1}). \tag{24.3}$$

The Equation (24.3) is an approximating sum of the area. The difference $(x_i - x_{i-1})$ in Equation 24.3 is the length of the *ith* subinterval in the partition and is sometimes denoted by the capital Greek letter Δx_i, "delta x sub i". Whether or not the partition is uniform, $\Delta x_i = (x_i - x_{i-1})$. Here, the partition is uniform so all the subintervals have the same length, so Δx_i is always $\frac{b-a}{n}$, in this case, $\frac{4-1}{6} = 0.5$.

To finish off this problem we will use $(x_i - x_{i-1}) = 0.5$, and compute the values of $\sqrt{x_{i-1}}$ for $i = 1, 2, 3, 4, 5, 6$. That is, for $x_0 = 1$, $x_1 = 1.5$, $x_2 = 2.0$, $x_3 = 2.5$, $x_4 = 3.0$, $x_5 = 3.5$, $x_6 = 4.0$. So, now

$$\begin{aligned} A &\approx (\sqrt{1})0.5 + (\sqrt{1.5})0.5 + \ldots + (\sqrt{3.5})0.5 \\ &\approx 4.411488\ldots. \end{aligned}$$

What would happen if we were to make a "refinement" of the partition $\mathcal{P}_6[1,4]$ by inserting more subintervals? Say, instead of $n = 6$, we want to make $n = 12$, getting the partition $\mathcal{P}_{12}[1,4]$. What would rectangles in this new approximating sum look like? See Figure 24.3.

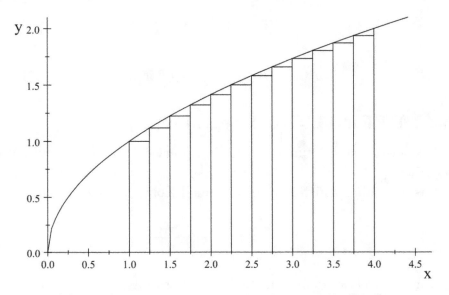

Figure 24.3 Approximating sum based on $\mathcal{P}_{12}[1,4]$.

Now the approximating sum would be

$$A \approx \sum_{i=1}^{12} \sqrt{x_{i-1}} \Delta x_i.$$

Where, this time, each Δx_i, $\frac{4-1}{12} = 0.25$. The approximating sum for $\mathcal{P}_{12}[1,4]$ is 4.5404..., which is closer to the area A, than the value obtained from $\mathcal{P}_6[1,4]$.

To see what the actual value is we find the area from our work with antiderivatives. If $A(x)$ is the area on the interval $[0,x]$, then for $t \in (0,x)$, area from t to x, would satisfy the following inequality,

$$\sqrt{t}(x - t) \leq A(x) - A(t) \leq \sqrt{x}(x - t).$$

Thus

$$\sqrt{t} \leq \frac{A(x) - A(t)}{x - t} \leq \sqrt{x},$$

and the limit as $t \to x$ yields

$$\frac{dA(x)}{dx} = \sqrt{x}.$$

So,

$$A(x) = \frac{2}{3}x^{3/2} + C,$$

and

$$A(4) - A(1) = \frac{14}{3} = 4.666....$$

24.3.4   DEFINITION: Integral of $f(x)$.

If $f(x)$ is a continuous function on the interval $[a, b]$ and n is a positive integer, then $\int_a^b f(x)dx$, read as "the integral of $f(x)$ with respect to x from a to b," is a number which may be obtained from the following limit

$$\lim_{n \to \infty} \sum_{i=1}^{n} f(x_{i-1})(\frac{b-a}{n}).$$

In other words, we define the definite integral as

$$\int_a^b f(x)dx = \lim_{n \to \infty} \sum_{i=1}^{n} f(x_{i-1})(\frac{b-a}{n}).$$

Here we used a uniform partition $\mathcal{P}_n[a, b]$ making the subinterval size $\frac{b-a}{n}$, and we used the left end point x_{i-1} of the *ith* subinterval to get the height of the *ith* approximating rectangle. We would get the same integral if we used the right end points of the subintervals or any other point in the subinterval to get the height of the approximating rectangles. Equivalently, we can define the definite integral as

$$\int_a^b f(x)dx = \lim_{n \to \infty, \, \Delta x_i \to 0} \sum_{i=1}^{n} f(\overline{x_i})\Delta x_i, \qquad (24.4)$$

where Δx_i is the length of the *ith* interval in any partition (not necessarily uniform) of $[a, b]$, and $\overline{x_i}$ is any number in the *ith* interval. We also assume that as n increases the length of the every subinterval approaches zero.

In the integral notation, $\int_a^b f(x)dx$, the expression $f(x)dx$ is called the *integrand of the integral.* The two numbers a and b are called the *lower limit* and *upper limit*, respectively. This is an unfortunate designation since these numbers are actually not limits at all, but they are the left and right endpoints of the interval over which the function $f(x)$ is being integrated.

We will want to see what happens when we write an integral as the sum of two integrals.

24.3.5   Problem

If $f(x)$ is a continuous function on the interval $[a, b]$ and t and u are two numbers in the interval $[a, b]$ such that $a \leq t < u \leq b$, prove that

$$\int_a^u f(x)dx = \int_a^t f(x)dx + \int_t^u f(x)dx. \tag{24.5}$$

Solution

We will use the following property of sums. If m is a positive integer and k is a positive integer such that $0 < k < m$, then

$$\sum_{i=1}^{m} a_i = \sum_{i=1}^{k} a_i + \sum_{i=k+1}^{m} a_i. \tag{24.6}$$

If we let $\mathcal{P}_k[a, t]$ be a partition of $[a, t]$, and $\mathcal{P}_{m-k}[t, u]$ be a partition of $[t, u]$, then $\mathcal{P}_k[a, t] \cup \mathcal{P}_{m-k}[t, u] = \mathcal{P}_m[a, u]$. Therefore,

$$\sum_{i=1}^{m} f(\overline{x_i})\Delta x_i = \sum_{i=1}^{k} f(\overline{x_i})\Delta x_i + \sum_{i=k+1}^{m} f(\overline{x_i})\Delta x_i. \tag{24.7}$$

The limits of these sums as $m \to \infty$, $k \to \infty$, and $\Delta x_i \to 0$ will be the integrals in Equation 24.5. See Figure 24.4.

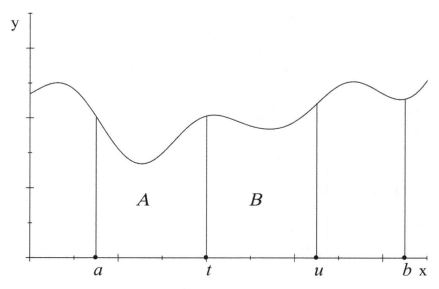

Figure 24.4 $A = \int_a^t f(x)dx$ and $B = \int_t^u f(x)dx$.

Two consequences of Equation (24.5) are

$$\int_a^t f(x)dx = \int_a^u f(x)dx - \int_t^u f(x)dx, \text{ and} \qquad (24.8)$$

$$\int_u^t f(x)dx = -\int_t^u f(x)dx. \qquad (24.9)$$

We leave the proofs of these as problems in the exercises. See Problem 2 in Exercise 24.

24.3.6 Definition: The integral as a function of its upper limit and lower limits.

Let a and b be fixed numbers with $a < b$, and $f(x)$ be a given continuous function on an interval $[a, b]$. Now, let t be any number, $a < t \le b$, then for the given number a and the given function $f(x)$, the integral $\int_a^t f(x)dx$ is a number that that depends on t. Define $F(t)$ as

$$F(t) = \int_a^t f(x)dx. \qquad (24.10)$$

We note that the integral is a function of the (variable) upper limit, t, and not of the variable x that runs from a to t. For the integrand $f(x)dx$, the variable x (that runs from a to t) is called the "dummy variable". Hey, I'm not making this up!

For a fixed continuous function $f(x)$ on a given interval $[a, b]$, if r and t are any two numbers in the segment (a, b), then the integral $\int_r^t f(x)dx$ is a function of both r and t, hence

$$F(r, t) = \int_r^t f(x)dx. \qquad (24.11)$$

Equations (24.10) and (24.11) are telling us that integrals represent quantities (area, volume, density, work, etc.) that can be expressed as functions of variables. And that these functions can be used in any mathematical formulas. We will be able, for example, to add, subtract, multiply and divide integrals. In particular, we can use the definition of derivatives to differentiate an integral. We can even use the definition of integrals to get an integral of an integral. The differentiation of an integral yields an important result which we will discuss later in this chapter, but first, let us introduce one more useful fact about integrals.

24.4   The mean value theorem for integrals

An ancient problem in Geometry was this: "Can we square the circle?" What this meant was: Given an area (such as the area of a circle) could we construct a rectangle whose area is the same as the given area?

24.4.1   Problem

Square the circle whose radius is r. That is, given a circle whose radius is r, find a square whose area is the same as the area of the circle.

Solution

Draw a circle whose radius is r. Draw a square whose sides are $r\sqrt{\pi}$. The area of the circle is πr^2, and the area of the square is also πr^2.

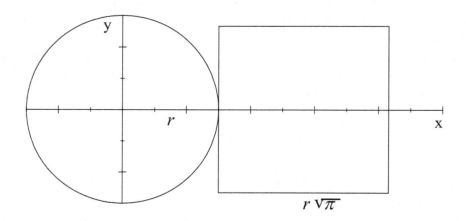

Figure 24.5 Circle and square with the same area.

Of course, this is not what was meant by the classic problem of squaring the circle, which required a construction of π using only Euclidean tools: an unmarked straight edge and a collapsible compass.   But it illustrates the idea of being able to replace an integral by a rectangle of the same area.

When the integral $\int_a^b f(x)dx$ represents the area under a curve over the interval $[a,b]$, "can we square the integral"? That is, can we find a number h such that the area defined by the integral is the same as the rectangle with base of length $b-a$ and height h? Or, $\int_a^b f(x)dx = h \times (b-a)$?

24.4.2   Example

If the integral $\int_a^b f(x)dx$ is the number A, find a rectangle whose base is the interval $[a,b]$ and whose area is A.

Answer

We simply divide A by $b-a$, getting a height of $h = \frac{A}{b-a}$. Thus, $A = h \times (b-a)$. The rectangle whose dimensions are $b-a$ and h will have area A.

If the number h is the ordinate of some point of the graph of $y = f(x)$, then h is called the *average height* of the function $f(x)$ over the interval $[a, b]$.

24.4.3 Problem

For the number h in the previous problem, show that h cannot be greater than the maximum value for $f(x)$.

Solution

Suppose that the graph of $y = f(x)$ has a maximum value, M, on the interval $[a, b]$. In Figure 24.6, the line $y = M$ passes through the maximum value of $f(x)$. Then for all $x \in [a, b]$, $|f(x)| \leq M$. So the area of the rectangle is $M \times (b - a)$, which is greater than $\int_a^b f(x)dx$. But $\int_a^b f(x)dx = h \times (b - a)$, so $h \leq M$. We leave the problem of proving that h cannot be less than the minimum value of $f(x)$ as homework Problem 4 in Exercise 24.

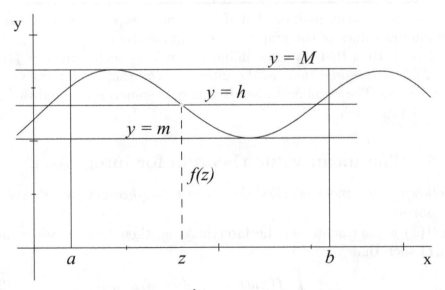

Figure 24.6. $\int_a^b f(x)dx = h \times (b - a)$.

24.4.4 Question

Does the function $f(x)$ actually "take on" the value h at some point in the interval $[a, b]$? In other words, if a function is continuous on an interval $[a, b]$, and if $f(x)$ has a maximum $y = M$ and a minimum $y = m$ on $[a, b]$ and if h is any number between M and m, does there exist a number z in $[a, b]$ such that $f(z) = h$?

Answer

Intuitively, this seems to be true and was always assumed to be true for many years after the invention of calculus. But the axiom needed to prove it was not developed until the beginning of the twentieth century. At that time many mathematicians proposed various axioms that were needed to answer this question and several other knotty questions about real variables and continuous and differentiable functions. In the next chapter, we will study one version of these axioms and use it to prove some theorems regarding limits, continuity and differentiable functions.

Let us temporarily assume that if h is a number between the maximum and minimum values of the graph of a continuous function $y = f(x)$, on the interval $[a, b]$, then there is some number z in $[a, b]$ such that $h = f(z)$. In Figure 24.6, we assume that, in the interval $[a, b]$, there is a number z such that $f(z) = h$. The number h discussed here is sometimes called the "mean value of the function".

24.4.5 The mean value theorem for integrals.

The following statement is called the *mean value theorem for integrals*.

Theorem

If $f(x)$ is continuous on the interval $[a, b]$, then there is some number $z \in (a, b)$ such that

$$\int_a^b f(x)dx = f(z)(b - a). \tag{24.12}$$

Proof.

Let $A = \int_a^b f(x)dx$. If h is the number $\frac{A}{b-a}$, then h is between the maximum and the minimum value of $f(x)$ on $[a, b]$. In the next chapter, we will prove that a continuous function on an interval takes on all the values in its range. Thus, $f(x)$ must be h for some number $x = z$ in $[a, b]$.

24.5   The derivative of the integral

We want to show that the integral is an antiderivative of f.

24.5.1   Problem

If $f(x)$ is a continuous function on the interval $[a, t]$, and $F(t)$ is the function

$$F(t) = \int_a^t f(x)dx,$$

find the derivative $\frac{dF(t)}{dt}$.

Solution

For any number r in the interval $[a, t]$, write

$$\begin{aligned}
F(r) &= \int_a^r f(x)dx, \text{ and}\\
F(t) &= \int_a^t f(x)dx.
\end{aligned}$$

Then

$$F(t) - F(r) = \int_a^t f(x)dx - \int_a^r f(x)dx.$$

By Equation (24.9), we can rewrite this as

$$F(t) - F(r) = \int_a^t f(x)dx + \int_r^a f(x)dx,$$

and by Equation (24.5)

$$F(t) - F(r) = \int_r^t f(x)dx. \tag{24.13}$$

Now by the mean value theorem, Equation (24.12), there is some number z, between r and t such that

$$\int_r^t f(x)dx = f(z)(t - r), \tag{24.14}$$

then by Equations (24.13), and (24.14),

$$\lim_{r \to t} \frac{F(t) - F(r)}{t - r} = \lim_{r \to t} f(z), \text{ or}$$
$$\frac{dF(t)}{dt} = f(t).$$

This means that

$$\frac{d}{dt} \int_a^t f(x)dx = f(t).$$

Thus, the integral $\int_a^t f(x)dx$, is an antiderivative of $f(t)$ on the interval $[a, t]$.. This result (that the definite integral is an antiderivative) is called the *Fundamental Theorem of Calculus*.

24.6 Exercise 24 HOMEWORK

1. When an object is moved some distance, s, in a straight line by a force F, then we say some work W, has been done, and the formula is

$$W = F \times s.$$

We will be consider the force to be weight. Thus, if the weight varies over the distance, s, then the force F is a function of s, that is, $F = F(s)$, and the work is $W = W(s)$. Suppose a force $F(s)$ acts over an interval in which $s \in [a, b]$. Let $\mathcal{P}_n[a, b]$ be a partition whose ith subinterval is of length $(s_i - s_{i-1})$ and assume that the force over the ith subinterval is a constant $F(s_{i-1})$, so that the work done over that interval is $F(s_{i-1}) \times (s_i - s_{i-1})$. An approximation of the work done over the whole interval $[a, b]$ is

$$W \approx \sum_{i=1}^{n} F(s_{i-1})(s_i - s_{i-1}).$$

The exact work would be the integral

$$W = \int_a^b F(s)ds.$$

(a) A 50 *lb* bag of sand is pulled up in a straight line to the top of a building $200.ft$ high, but the sand is leaking out of the bag at a rate of $\frac{1}{7}$ *lb* a pound per foot, how much work was done in lifting this bag?

(b) How much work would have been done if the bag did not leak?

2. Using the equation

$$\int_a^u f(x)dx = \int_a^t f(x)dx + \int_t^u f(x)dx,$$

prove

(a) $\int_a^t f(x)dx = \int_a^u f(x)dx - \int_t^u f(t)dt.$

(b) $\int_u^t f(x)dx = -\int_t^u f(x)dx.$

(c) $\int_a^a f(x)dx = 0.$

3. Let n be an integer and i be a positive integer less than or equal to n. If, a_i and b_i are numbers such that for all i, $a_i \le b$, then the following equation is true.

(a)

$$\sum_{i=1}^{n} a_i \le \sum_{i=1}^{n} b_i.$$

Prove that if $f(x) \le g(x)$ for all $x \in [a, b]$, then

$$\int_a^b f(x)dx \le \int_a^b g(x)dx.$$

(b) Prove that if $f(x) = c$, a constant on the interval $[a, b]$, then

$$\int_a^b f(x)dx = c \times (b - a).$$

4. Given that the integral $\int_a^b f(x)dx = A$, let $h = \frac{A}{b-a}$. If m is the minimum of the graph of $y = f(x)$ on $[a, b]$, prove that $h \geq m$.

5. Given the following properties of sums are true.

$$\sum_{i=1}^n a_i + \sum_{i=1}^n b_i = \sum_{i=1}^n (a_i + b_i),$$

and, for any number y that does not depend on i,

$$y \sum_{i=1}^n a_i = \sum_{i-1}^n ya_i.$$

Prove

(a) $\int_a^b f(x)dx + \int_a^b g(x)dx = \int_a^b (f(x) + g(x))dx$, where each of f and g is a continuous function on $[a, b]$.

(b) $y \int_a^b f(x)dx = \int_a^b yf(x)dx$, where y is any number that does not depend on x.

6. Prove

(a) For any integer n, the following dummy variable equation for sums is true.

$$\sum_{i=1}^n i^2 = \sum_{j=1}^n j^2.$$

(b) For any interval $[a, b]$, the following dummy variable theorem for integrals is true.

$$\int_a^b f(x)dx = \int_a^b f(t)dt.$$

7. Let $x > 0$, and let A be the region bound by the graph of $y = x^2$ and $y = 8 - 2x$.

   (a) Sketch the two graphs and shade the region A.

   (b) Sketch a partition $\mathcal{P}_4[0, 2]$ and construct a set of rectangles, the sum of whose areas approximate the area of the region A.

   (c) Find the integral which gives you the exact area of A.

8. For each of the following integrals, find their derivatives by the definition of derivative.

   (a) Given a is a constant, find $\frac{dF(t)}{dt}$ if

   $$F(t) = \int_a^{t^2} f(x)dx.$$

   (b) Given a is a constant, and $g(t)$ is a differentiable function, find $\frac{dF(t)}{dt}$, if

   $$F(t) = \int_a^{g(t)} f(x)dx.$$

   (c) Find $\frac{dF(t)}{dt}$, if

   $$F(t) = \int_{\pi/2}^{t^2+t} \sin(x)dx.$$

Chapter 25

The S_1, S_2 axiom

25.1 Questions about completeness of the real numbers

- What is a proof in mathematics?

- The S_1, S_2 axiom.

- Questions about continuity.

- Rolle's Theorem.

- The mean value theorem for derivatives.

25.2 What is a proof in mathematics?

To answer the question, "What is a proof in mathematics?" we need to answer another question: "What is mathematics?" We have been studying calculus, which is one branch of mathematics, and it depends on the real number system. Another branch of mathematics, such as Number Theory, depends upon the positive integers. Algebra and Group Theory depend on rules for combining elements of sets. Complex analysis generalizes calculus and permits the use of imaginary numbers. Topology depends upon sets and ways to combine subsets. Probability depends upon random processes. Geometry, the fountainhead of practically all of mathematics, depends upon

267

assumptions about points, lines, and space.   And there is not just one kind
of geometry.   Euclidean geometry depends upon one set of assumptions
about parallel lines and Lobachevskian geometry depends upon another set
of assumptions.

Occasionally, these branches are combined into still other, newer branches.
But every one of these have one thing in common, and that is, each one has
its own set of assumptions. A statement that is true in one branch may not
be true, or even meaningful, in another branch.   For example in the complex
number system the equation $x^2 = -4$ can be solved.   It has two solutions
$x = 2i$, and $x = -2i$, where i is defined as $\sqrt{-1}$.   But in the real number
system the same equation, $x^2 = -4$ has no solution.   We can prove that if
x is any real number, then by the axioms of the real number system it must
be true that $x^2 \geq 0$.

But this does not say that the complex number system is right and the
real number system is wrong, nor does it say the real number system is
right and the complex number system is wrong.   Suppose we compare two
different board games, chess and checkers.   These games have different rules;
in fact, the rules define the game.   A move that is legal in one game is not
necessarily legal in the other, but that doesn't make one move right and the
other wrong.   If we want to check whether or not a move is right, we simply
ask the question "Is this move consistent with the rules of this game?"

What happens if we really like a move and want to be able to play it?
Then we change the rules to allow it.   Of course, what we have done is
created a new game in which the previously disallowed move is now okay.
Changing rules or adding new rules occurs frequently in professional sports.

This is the nature of "truth" in a mathematical statement. If a statement
is logically consistent with the assumptions (the axioms) of the mathematical
system, then we say the statement is true.   But how do we know the statement is consistent with the axioms? Ah, that requires a proof. Namely, we
must show exactly how the statement under consideration can be derived,
using logical steps from the axioms and the definitions and the undefined
terms that we are assuming for this particular mathematical system.

It may take years before we can come up with a proof of a statement
that we believe to be true.   In 1637, Pierre de Fermat asserted his famous
statement, which we now call Fermat's Last Theorem.   He claimed that he
had a proof but he never revealed it.   The statement was believed to be true,
and many mathematicians tried to prove it, but it remained unproved for
356 years. Finally, in 1993, Andrew Wiles was able to prove Fermat's Last

Theorem by bringing together an elaborate collection of new statements that he derived himself, plus other related statements previously proved over the years since Fermat. Wiles' proof required a book of over 200 pages.

With every true statement in a mathematical system there is a corresponding false statement. If you are having difficulty proving the suspectedly true statement, it is sometimes, but not always, easier to try to disprove its denial. This method is called *a proof by indirect argument.*

One version of a proof by indirect argument is the method of proof called *mathematical induction.* An example is, you notice something strange about the sum of the odd numbers. The sum of the first 2 odd numbers is 2^2, $1 + 3 = 4$. The sum of the first 3 odd numbers is 3^2, $1 + 3 + 5 = 9$. The sum of the first 4 odd numbers is 4^2, $1 + 3 + 5 + 7 = 16$. And you wonder is it always true that the sum of the first n odd numbers is the square of n? You could keep verifying this equation for one integer after another until you die and you would still not have proved this is always true. But you could use an indirect proof as follows. Suppose there is some number n where this statement is no longer true. Let's call this and other such numbers "bad" numbers. Let B be the set of all such bad numbers. We know that $1, 2, 3,$ and 4 are NOT BAD. So all of the bad numbers must be beyond 4. In other words, if $x \in B$, then $x > 4$. This means that B is *bounded below* and that 4 is one of its lower bounds, as are $1, 2, 3,$ and any other integer for which you have made the calculation to verify the equation.

There is an axiom for the real numbers called the well-ordering principle which says that any set of positive integers bounded below has to have a smallest element. Let m be the smallest element in the set B. That is, m is the smallest bad number. Therefore $m - 1$ is not a member of B. This means that the sum of the first $m - 1$ odd numbers *is* $(m - 1)^2$, but the sum of the first m odd numbers *is not* m^2. Therefore,

$$1 + 3 + 5 + \ ... \ + 2m - 3 \ = \ (m - 1)^2, \text{ but} \qquad (25.1)$$
$$1 + 3 + 5 + \ ... \ + 2m - 3 + 2m - 1 \ \neq \ m^2 \qquad (25.2)$$

If you subtract an equation from an inequality, you will get an inequality. Subtract Equation (25.1) from the Inequality (25.2), so

$$2m - 1 \neq 2m - 1.$$

This is a contradiction, therefore m does not exist. That is, there is no first element in B.

In this chapter we will introduce an axiom for the real number system which is actually a generalization of the well ordering principle.

25.3   The S_1, S_2 axiom

Some time during the two decades from 1890 to 1910, mathematicians began to notice that there was something incomplete about the set of axioms that defined the real numbers. They had been basing the calculus on a set of axioms involving several intuitive ideas, but there were some problems they could not solve. For example, they could not prove that a set bounded above had to have a right-most point just because it has a *Least Upper Bound* (LUB) or that a set bounded below necessarily had to have a left-most point, even if it had a *Greatest Lower Bound* (GLB). To fix this problem, and make calculus more rigorous some mathematicians in the late 1890's, started to suggest the addition of a new axiom in the real number system. The generic name for this new axiom is "the completeness axiom". One version of this axiom, developed around 1888 by Richard Dedekind became known as the "Dedekind Cut Axiom" and is the one we used in Dr. Moore's calculus class, where it was known as the "S_1, S_2 axiom".

25.3.1   Axiom S_1,S_2

If S_1 and S_2 are two point sets on the x-axis such that every point of the x-axis belongs either to S_1 or to S_2 and every point of S_2 is to the right of every point of S_1, then there is either a right-most point of S_1 or a left-most point of S_2.

See Figure 25.1

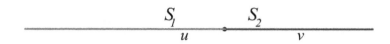

Figure 25.1 If $u \in S_1$ and $v \in S_2$ then $u < v$.

25.3.2 Problem

Prove that it is not possible for both S_1 to have a right most-point and S_2 have a left-most point.

Solution

Suppose that S_1 has a right-most point, z_1, and S_2 has a left-most point, z_2, then they are two distinct points; that is, $z_1 \neq z_2$. Therefore there is a point w between them. This point w cannot belong to S_1 because it is to the right of the right-most point in S_1 and w cannot belong to S_2 because it is to the left of the left-most point of S_2. This contradicts the requirement that every point of the x-axis must belong to either S_1 or S_2.

25.3.3 Problem

Prove that it is not possible for a point to belong to both sets S_1 and S_2.

Solution

Suppose a point x of S_2 belongs to S_1, then it would be to the left of every point of S_2 including itself.

25.3.4 Problem

If F is the graph of a continuous function on the interval $[a, b]$ and L is a horizontal line containing more than one point of F. Prove that there is a left-most point of F on L.

Solution

Let K be the set of all points of F that are on the line L.   Suppose

1. S_1 is a point-set such that p belongs to it if, and only if, p is a point of the x-axis such that *every* point of K is to the right of the vertical line through p..

2. S_2 is a point set such that p belongs to it if, and only if, p is a point of the x-axis such that *there is some* point of K that is either to the left of the vertical line through p or on the vertical line through p.

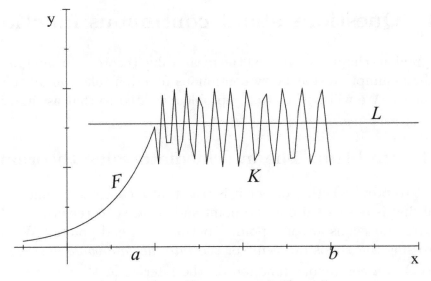

Figure 25.2 K = Points of F on L.

Every point of the x-axis has to be either in S_1 or in S_2. Every point of S_2 is to the right of every point of S_1. Therefore by the axiom either S_1 has a right-most point or S_2 has a left-most point. Suppose z is the left most point of S_2, then either there is some point of K to the left of the vertical line through z or on the vertical line through z. But no point of K can be to the left of z, otherwise there would be a point of S_2 to the left of the left-most point of S_2. So z is on a vertical line passing through a point of K, which is the left-most point of K.

Suppose S_2 does not have a left most-point, but that S_1 has a right-most point, z. Here we will get a contradiction as follows. Every point of K is to the right of z, therefore the point $P = (z, f(z))$ cannot be a point of K. Hence P must be either above L or below L. If P is above L, then by continuity there are two vertical lines h and k, with P between them such that F is above L for all points between h and k. But all the points of F to the right of z and to the left of the vertical line k are in S_1. This means that there are points of S_1 to the right of the right-most point of S_1. This is a contradiction. If we consider the case in which P is below L, we also get a contradiction by a similar argument. So S_2 has a left most point and thus K has a left most point.

25.4   Questions about continuous functions

In the previous chapter we proved the mean value theorem for integrals based upon the assumption that every continuous function takes on all the values in its range. We will now discuss a problem related to that assumption.

25.4.1   Problem: The intermediate value theorem

We want to prove whether or not it is true that a continuous function on an interval that is positive at one end point and negative at the other end point, must cross the x-axis at some point z between the end points. We call this the "crossing the x-axis problem" or the *intermediate value theorem*.

If $f(x)$ is a continuous function on the interval $[a, b]$ and $f(a) < 0$ and $f(b) > 0$, does there exist a number $z \in (a, b)$ such that $f(z) = 0$?

Solution

Let F be the graph of the equation $y = f(x)$, part of which is shown in Figure 25.3.

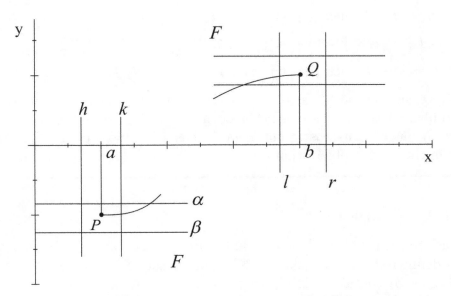

Figure 25.3 Graph with $f(a) < 0$ and $f(b) > 0$.

Since F is continuous on the interval, it is continuous at the end points $P = (a, f(a))$ and $Q = (b, f(b))$. We have drawn two horizontal lines α and β with P between them and found two vertical lines, h and k with P between them and see that every point of the F between h and k *is* also between α and β. This shows that F is below the x-axis in an interval containing a. Similarly, continuity at Q, means that F is above the x-axis in an interval containing b.

We can use this information to define two sets S_1, and S_2 on the x-axis. Let S_1 be the set of all numbers, t, on the x-axis for which it is true that there exist no points of F to the left of t which are above the x-axis. The set S_1 has some points in it, namely all the points to the left of a and all of the points between a and the vertical line k.

We define the set S_2 to be the set of points that satisfy the denial of the requirements for points in S_1. That is, we define the set S_2 to be the set of all points, t, on the x-axis for which it is true that there exists some point of F to the left of t that is above the x-axis. The set S_2 has points in it, namely all of the points to the right of b, and all of the points between b and the vertical line l shown in Figure 25.3.

Here is a summary of the definition of these two sets. Let R be the set of all real numbers, then

1. $S_1 = \{t \in R$ such that either $t < a,$

   or $t \geq a$, and for all $x < t$, $f(x) \leq 0\}$.

2. $S_2 = \{t \in R$ such that for some number $x < t$, $f(x) > 0\}$.

Every number on the x-axis is either in the set S_1 or in the set S_2. Also no number of the x-axis can be in both S_1 and S_2. By the S_1, S_2 axiom either S_1 has a right-most point z_1 or S_2 has a left-most point z_2. We will show in the next problem that either $f(z_1) = 0$ or $f(z_2) = 0$.

25.4.2 Problem

Go back to the problem illustrated in Figure 25.3, and assume the sets S_1 and S_2 as defined above. By the S_1, S_2 axiom, either S_1 contains a right-most point z_1 or S_2 contains a left-most point z_2.

Part 1. Start with the assumption that S_2 has a left most point, z_2 then show that $f(z_2) = 0$.

Solution

Assume that S_2 has a left-most point, z_2. This point cannot be to the left of a because those points are in S_1. Thus, z_2 must be somewhere in the interior of the interval $[a, b]$, because $b \in S_2$ and so are the points in the segment (a, b) that are between b and the vertical line l. Now, suppose that $f(z_2) \neq 0$. Then either $f(z_2) > 0$ or $f(z_2) < 0$.

1. Case 1. $f(z_2) > 0$. By continuity at z_2 if β_1 is a horizontal line whose equation is $y = \frac{1}{2}f(z_2)$, then there is a vertical line to the left of z_2 with points of F above β_1, which means that these points are also where F is above the x-axis making them all members of S_2. That is, they are points of S_2 to the left of the left-most point of S_2. This is a contradiction.

2. Case 2. $f(z_2) < 0$. For z_2 to be the left-most point of S_2, all the
   points to the left of z_2 are is S_1. But since $f(z_2) < 0$, then there is
   an interval containing z_2 such that the graph F is below the x-axis for
   every point in that interval. This means that there is no point, x, to
   the left of z_2 such that $f(x) > 0$. Hence, z_2 is not a point of S_2. This
   is a contradiction.

These two contradictions prove that $f(z_2) = 0$, if z_2 is the left-most point
of S_2.

25.4.3   Problem

Part 2. Assume that S_1 has a right-most point, z_1, and prove that $f(z_1) = 0$.

Solution

If z_1 is the right-most point of S_1 it cannot be to the right of b, because
b and all the numbers of $[a, b]$ between the vertical lines l and r are points
of S_2. Also z_1 is interior to the interval $[a, b]$ because there are points of
S_1 between the vertical lines h and k. Suppose $f(z_1) \neq 0$. Then either
$f(z_1) > 0$ or $f(z_1) < 0$.

1. Case 1. $f(z_1) > 0$. Then by continuity at z_1 there are vertical lines
   on either side of $(z_1, f(z_1))$ such that the graph F is above the x-axis,
   hence there are points of F to the left of z_1 above the x-axis, meaning
   that z_1 is in S_2 not S_1. This is a contradiction.

2. Case 2. $f(z_1) < 0$. By continuity at z_1 there are vertical lines that
   define a segment containing z_1 such that for all x in that segment,
   $f(x) < 0$. This means that there are points of S_1 to the right of z_1.
   This is a contradiction.

These two contradictions prove that $f(z_1) = 0$, if z_1 is the right-most
point of S_1.

We have just solved the "crossing the x-axis problem". Now try to solve
the following problem.

25.4.4   Problem

Suppose $f(x)$ is continuous on the interval $[a, b]$, and there exists numbers M and m such that $m \leq f(x) \leq M$ for all $x \in [a, b]$. If h is any number between M and m then there is a number z in the interval $[a, b]$ such that $f(z) = h$.

Solution

Denote a maximum point of $f(x)$ as $P = (x_0, M)$ and a minimum point $Q = (x_1, m)$. Let $m < h < M$. (If h is either M or m, then the problem is solved since z would be either x_0 or x_1.) Define a function $g(x) = f(x) - h$, then $g(x_0) = M - h > 0$ and $g(x_1) = m - h < 0$. Therefore, $g(x)$ is a continuous function that is negative at one point and positive at another point, so it has a point on the x-axis, namely, there exists a number z in $[a, b]$ such that $g(z) = 0$. This proves that $f(z) = h$.

This is a generalization of the "crossing the axis problem". What is says is that if a continuous graph on an interval has a point below the horizontal line $y = h$ and another point in that interval above the same line then it must have a point *on* that line in the same interval.

25.4.5   Problem

If $f(x)$ is a function on the interval $[a, b]$ such that the derivative of $f(x)$ is continuous on (a, b) and $f'(a) > 0$ and $f'(b) < 0$, show that there is some point z in (a, b) such that $f'(z) = 0$.

Solution

This is simply an example of the crossing the axis problem. See Figure 25.4(a) and 25.4(b).

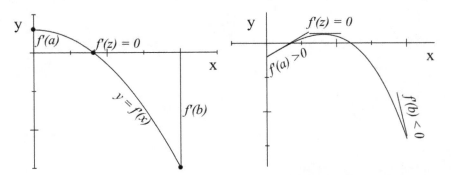

Figure 25.4(a) Graph of $f'(x)$.   Figure 25.4(b). Graph of $f(x)$.

25.5   Rolle's Theorem

In the preceding section we saw that a function that has a continuous derivative at all of its points and has a positive derivative at one end point and
a negative derivative at the other end point must have a derivative equal to
zero at some point between the two end points.   This is useful for finding
the highest or lowest point of the graph.

In turns out that we can find a point at which a function has a zero
derivative if we know only that the function has a continuous derivative on
some interval and the function is zero at its end points.   This was discovered
in 1690 by M. Rolle, and is known today as Rolle's Theorem.   We will now
state it as a problem for you to solve.

25.5.1   Problem

If G is the graph of a function $y = g(x)$ whose derivative is continuous on
the interval $[a, b]$, and such that $g(a) = 0$ and $g(b) = 0$, prove that there is
some number z in the segment (a, b), such that $g'(z) = 0$.

Solution

_____

Figure 25.5 shows the given conditions that $g(a) = g(b) = 0$.   One of
the steps in our proof requires that we consider a number $t \in (a, b)$ at which

$g'(t) > 0$. Such a possible point is also shown in the graph.

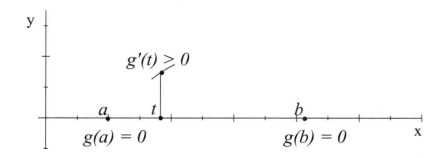

Figure 25.5 Showing that $g(a) = g(b) = 0$.

Let G be the graph of a function $g(x)$ whose derivative is continuous on $[a, b]$. Now, let t be any point in the segment (a, b). Then $g'(t)$ exists, and is either $g'(t) > 0$, or $g'(t) = 0$, or $g'(t) < 0$.

If $g'(t) = 0$, then the problem is solved because t is a number z that we want to show exists. If $g'(t) > 0$, then either $g'(x) > 0$ for all x in (a, b) to the left of t, or there is some $x_0 \in (a, b)$ to the left of t such that $g'(x_0) \leq 0$. Thus, either $g'(x_0) = 0$, and the problem is solved because x_0 is a number z or $g'(x_0) < 0$, which with $g'(t) > 0$, makes $g'(z) = 0$, for some z between x_0 and t, by the axis crossing problem.

Now, either there is some number x_1 between t and b such that $g'(x_1) \leq 0$ which is sufficient to conclude that there exists a number z between t and x_1 such that $g'(z) = 0$. Or, $g'(x) > 0$ for all x in $[t, b]$ which implies that the graph G is increasing throughout the entire interval $[a, b]$. Thus, for any two numbers t and r in (a, b), with $r > t$, we have $g(r) > g(t) > 0$. Let α_1 be a horizontal line with equation $y_0 = g(r)/2$ The point B of G is $(b, 0)$, and every pair of vertical lines with B between them will have a point of G above α_1. This means that G is not continuous at B, contrary to the given that G is continuous on $[a, b]$. Therefore $g'(x_1) < 0$ for some x_1 in $[t, b]$. This means that there is some number z between t and x_1 such that $g'(z) = 0$.

We started this proof assuming the case in which $g'(t) > 0$, but had we assumed that $g'(t) < 0$, we could use a similar argument to prove that $g'(z) = 0$ for some z in (a, b)

A stronger version of Rolle's theorem can be proved. Its statement is as follows: If G is the graph of a differentiable function $g(x)$ defined on the interval $[a, b]$ and if $g(a) = g(b) = 0$, then there is a number $z \in (a, b)$, such that $g'(z) = 0$. In this case we are not assuming the derivative of g is continuous on $[a, b]$, just that it exists there.

25.6 Mean value theorem for derivatives

Rolle's Theorem leads us to another theorem which says that if $y = g(x)$ is a continuously differentiable function on an interval $[a, b]$, then there is some number z in the segment (a, b) such that

$$g'(z) = \frac{g(b) - g(a)}{b - a} \tag{25.3}$$

which is called the *mean value theorem for derivatives*. The proof of this uses the idea that a new function $f(x)$ can be constructed by adding and subtracting certain things that would make $f(a) = f(b) = 0$. This is a fun problem to work on. If you want a hint, try something like the following. Think of a number m that would make the definition $f(x) = g(x) - g(a) - m(x - a)$ a useful function.

A generalization of Equation (25.3) is as follows: If $f(x)$ and $g(x)$ are two continuously differentiable functions on the interval $[a, b]$, with $f(a) = g(a)$ and $f(b) = g(b)$, then there is some number z in the segment (a, b), such that $f'(z) = g'(z)$. This is also fun to work on. Think about it.

NOTE: The set of exercise problems and solutions for this chapter (Chapter 25) is the empty set.

Appendix

Solutions to Exercises

NOTE TO THE STUDENT:

Do not look at the answer to any homework problem that you are having difficulty with until you have spent at least one-half hour on it. Set this problem aside for a while, perhaps overnight, when you come back to it you may find that you can readily solve it. Such is the manner of mathematical creativity.

The following pages show the solutions to Homework Problems

1. The next few pages reveal answers to problems you may still be working on.

2. You may wish to **wait** until you have solved these before going on.

3. Instead of reading the answers given in the following pages, you may use these pages as **HINTS** for solving the problem, here's how:

- Cover up the page with a sheet of paper.

- Slide the sheet down and expose one line at a time.

- Try to work out for yourself what the next line should be.

4. You can **check** your results against the solutions given here.

1. Rapid Sketching

Answers to Exercises 1

1. $y = -(x-1)(x-2)$. The points $(1,0)$ and $(2,0)$ are on the graph. If $x < 1$, then both $(x-1)$ and $(x-2)$ are negative. Their product is positive; so $-(x-1)(x-2) < 0$, hence the graph is below the x-axis. However, when $1 < x < 2$, the graph is above and when $x > 2$, below. See Exercise 1 Figure 1.

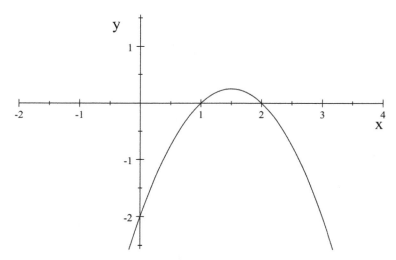

Exercise 1 Figure 1 $y = -(x-1)(x-2)$.

2. $y^2 = -(x-1)(x-2)$. Look back at Exercise 1 Figure 1. Where *that* graph is above the x-axis, *this graph* will have two branches because $-(x-1)(x-2)$ will have two square roots. But where the graph in Exercise 1 Figure 1 is below x-axis *this graph* does not exist. See Exercise 1, Figure. 2.

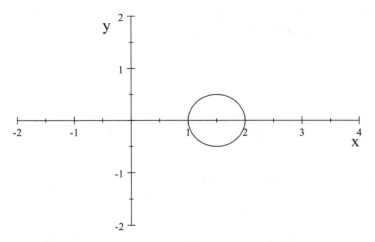

Exercise 1 Figure 2 $y^2 = -(x-1)(x-2)$.

3. $y = (x-1)^2(x-2)$.   The value of y is zero at $x = 1$ and $x = 2$, but when x is less than 1, $(x-1)^2(x-2)$ is negative because $x-2$ is negative and $(x-1)^2 > 0$. Also if $x > 1$ and $x < 2$, the product $(x-1)^2(x-2)$ is again negative. The value of y is positive only when $x > 2$. See Exercise 1, Figure 3.

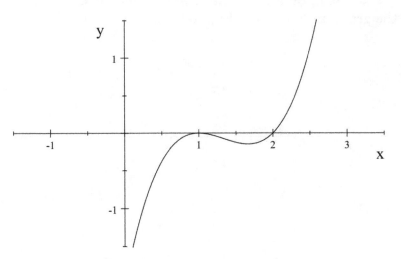

Exercise 1 Figure 3 $y = (x-1)^2(x-2)$.

4. $y^2 = (x-1)^2(x-2)$.   This equation is equivalent to the two square root equations $y = \sqrt{(x-1)^2(x-2)}$ and $y = -\sqrt{(x-1)^2(x-2)}$. Its

graph will exist only where the graph in Exercise 1 Figure 3 is not below the x-axis. This occurs at $x = 1$ or for all $x \geq 2$. See Exercise 1 Figure 4.

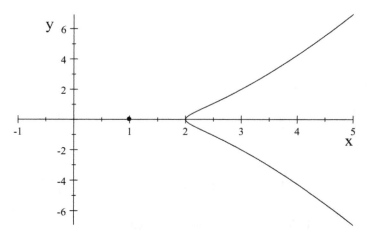

Exercise 1 Figure 4 $y^2 = (x-1)^2(x-2)$.

5. $y = x(x+2)(x-3)$. The graph cuts the x-axis at $x = 0$, $x = -2$, and $x = 3$. The graph is above the x-axis (that is, $y > 0$) when either $-2 < x < 0$, or $x > 3$. Also, $y < 0$ when $0 < x < 3$ or $x < -2$. See the graph in Exercise 1 Figure 5.

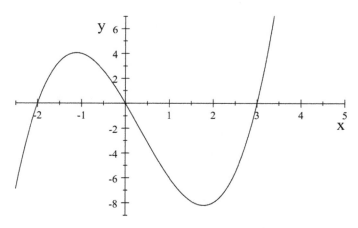

Exercise 1 Figure 5 $y = x(x+2)(x-3)$.

6. $y^2 = x(x+2)(x-3)$. This equation is equivalent to the two equations $y = \sqrt{x(x+2)(x-3)}$ and $y = -\sqrt{x(x+2)(x-3)}$; its graph is the

square root of the graph in Problem 5. We get the graph shown in Exercise 1 Figure 6.

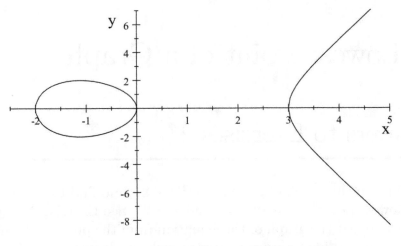

Exercise 1 Figure 6 $y^2 = x(x+2)(x-3)$.

2. Lowest Point of a Graph

Answers to Exercises 2

---

1. Sketch the graph of $y = (x-1)^2(x-2)$, and find the coordinates of its lowest point between $x = 1$ and $x = 2$. See Exercise 2 Figure 1 where we repeat the graph of this equation from the previous homework, here we have added a secant line through the lowest point Q and another point P.

The coordinates of P and Q are

$$
\begin{aligned}
P &= (p, (p-1)^2(p-2)), \\
Q &= (q, (q-1)^2(q-2)).
\end{aligned}
$$

The slope of the line PQ is

$$
m_{PQ} = \frac{(p-1)^2(p-2) - (q-1)^2(q-2)}{p-q}.
$$

When multiplied out, this expands to

$$
m_{PQ} = \frac{p^3 - 4p^2 + 5p - 2 - (q^3 - 4q^2 + 5q - 2)}{p-q}.
$$

After re-grouping like powers of p and q and canceling the -2 and $+2$, we get

$$
m_{PQ} = \frac{(p^3 - q^3) - 4(p^2 - q^2) + 5(p - q)}{p-q}.
$$

We want P to approach Q so as to get the tangent line at Q, but this would make $p - q$ be zero in the denominator. Fortunately, however,

289

each term in the numerator can be factored so that the numerator has a matching factor $(p - q)$.

$$m_{PQ} = \frac{(p - q)(p^2 + pq + q^2) - 4(p - q)(p + q) + 5(p - q)}{(p - q)}.$$

Before letting $p \to q$, we cancel out the common factor $(p - q)$ from both the numerator and denominator, leaving

$$m_{PQ} = p^2 + pq + q^2 - 4p - 4q + 5. \qquad \text{(I)}$$

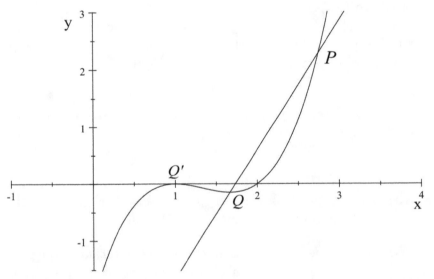

Exercise 2 Figure 1 $y = (x - 1)^2(x - 2)$ and the secant line PQ.

Now as $P \to Q$, the secant line PQ becomes the tangent line, and $p \to q$, so in Equation (I) $p^2 \to q^2$, $pq \to q^2$, and q^2 stays q^2. Similarly, $-4p \to -4q$, and 5 stays 5. Thus, the slope m_{PQ} becomes

$$m = 3q^2 - 8q + 5. \qquad \text{(II)}$$

where m is the slope of the tangent line to G at the point Q. If Q is the lowest point then the slope $m = 0$. Then from Equation (II) we have

$$3q^2 - 8q + 5 = 0.$$

And by the quadratic formula

$$q = \frac{8 \pm \sqrt{64 - 60}}{6}.$$

The two roots are $q = \frac{5}{3}$, or $q = 1$. This means there are two points with tangent lines that have zero slope. One of them, Q is the low point and it is between 1 and 2. So the abscissa of Q is $q = \frac{5}{3}$. To find the ordinate of Q, we go back to the equation $y = (x-1)^2(x-2)$ and substitute $\frac{5}{3}$ for x getting $(\frac{5}{3} - 1)^2(\frac{5}{3} - 2) = -\frac{4}{27}$. Thus, the coordinates of Q are $(\frac{5}{3}, -\frac{4}{27})$. If we substitute the other solution $q = 1$ for x in $y = (x-1)^2(x-2)$, we get $y = 0$ as the ordinate of the point $Q' = (1, 0)$. At this point, the tangent is also zero. It is the highest point of the graph for all $x < 2$. See Exercise 2, Figure 1.

2. Graph $y = x(x + 2)(x - 3)$ and find the lowest point between $x = 1$ and $x = 3$.

   (a) This graph is shown here in Exercise 2 Figure 2 with the points P, Q, Q' and the secant line PQ added.

   (b) To find the coordinates of the lowest point, Q of the graph between 1 and 3, let p be the abscissa of the point P and q be the abscissa of the point Q. Since P and Q are points on the graph whose equation is $y = x(x + 2)(x - 3)$, the coordinates of P and Q are

$$\begin{aligned} P &= (p, p(p+2)(p-3)), \\ Q &= (q, q(q+2)(q-3)). \end{aligned}$$

The slope of the secant line PQ is

$$m_{PQ} = \frac{p(p+2)(p-3) - q(q+2)(q-3)}{p - q}. \qquad \text{(III)}$$

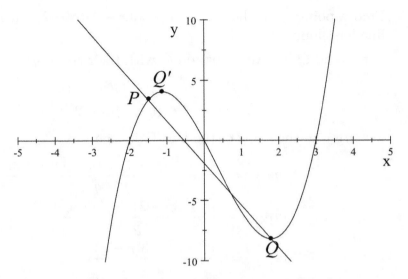

Exercise 2 Figure 2 Secant line PQ in Problem. 2.

If we try to let $p = q$ in Equation (III), we get the indeterminate form $\frac{0}{0}$. Let us first multiply out the numerator and regroup to get

$$m_{PQ} = \frac{(p^3 - p^2 - 6p) - (q^3 - q^2 - 6q)}{p - q}$$

$$m_{PQ} = \frac{(p^3 - q^3) - (p^2 - q^2) - 6(p - q)}{p - q}. \qquad \text{(IV)}$$

Factor the numerator, cancel the $(p - q)$'s, and simplify;

$$m_{PQ} = p^2 + pq + q^2 - (p + q) - 6.$$

Now we can let $p \to q$ and we find that the slope of the tangent line at Q is

$$m = 3q^2 - 2q - 6.$$

Setting $m = 0$ yields the two values of q

$$q = \frac{2 \pm \sqrt{76}}{6} = \frac{1 \pm \sqrt{19}}{3}.$$

Let Q' be the highest point of G between -2 and 0; it has abscissa $\frac{1 - \sqrt{19}}{3} \approx -1.12$. See Exercise 2, Figure 2. The lowest point, Q between 0 and 3 has abscissa $\frac{1 + \sqrt{19}}{3} \approx 1.79$.

3. Find a point Q on the graph of $y = (x-1)(x-2)$ where the tangent line has slope -1.

Let P and Q be two points on G with their coordinates

$$
\begin{aligned}
P &= (p,\ (p-1)(p-2)),\\
Q &= (q,\ (q-1)(q-2)).
\end{aligned}
$$

Let m_{PQ} be the slope of the line PQ, then

$$
\begin{aligned}
m_{PQ} &= \frac{(p-1)(p-2)-(q-1)(q-2)}{p-q}\\
m_{PQ} &= \frac{p^2-3p+2-(q^2-3q+2)}{p-q},\\
m_{PQ} &= \frac{p^2-q^2-3(p-q)}{p-q}.
\end{aligned}
$$

After factoring out $(p-q)$ and canceling it from the numerator and denominator, we can let $p \to q$, getting $m = 2q - 3$ as the slope of the tangent line at Q. To find what value of q makes the slope $= -1$, we set $m = -1$, or $2q - 3 = -1$. This yields $q = 1$. Thus the graph has slope -1 at the point $(1,0)$. See Exercise 2 Figure 3.

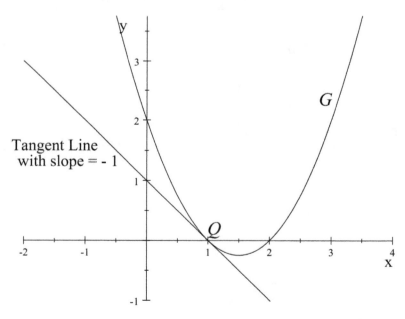

Exercise 2 Figure 3. Line with slope -1 tangent to G at $Q = (1,0)$.

4. For the given equation

$$p = \frac{\sqrt{r^2 + r + 2} - \sqrt{s^2 + s + 2}}{r - s}.$$

(a) We want to "rationalize" the numerator. Multiply the numerator and denominator by

$$\sqrt{r^2 + r + 2} + \sqrt{s^2 + s + 2},$$

getting

$$p = \frac{\left(\sqrt{r^2 + r + 2} - \sqrt{s^2 + s + 2}\right)\left(\sqrt{r^2 + r + 2} + \sqrt{s^2 + s + 2}\right)}{(r - s)\left(\sqrt{r^2 + r + 2} + \sqrt{s^2 + s + 2}\right)},$$

or

$$p = \frac{(r^2 + r + 2) - (s^2 + s + 2)}{(r - s)\left(\sqrt{r^2 + r + 2} + \sqrt{s^2 + s + 2}\right)}.$$

Grouping terms and factoring yields

$$p = \frac{(r - s)(r + s) + (r - s)}{(r - s)\left(\sqrt{r^2 + r + 2} + \sqrt{s^2 + s + 2}\right)}.$$

This simplifies to

$$p = \frac{r + s + 1}{\left(\sqrt{r^2 + r + 2} + \sqrt{s^2 + s + 2}\right)}.$$

(b) It is now safe to let $r \to s$, so the value of p when $r = s$ is

$$\frac{2s + 1}{2\sqrt{s^2 + s + 2}}.$$

5. $y^2 = x(x + 2)(x - 3)$

(a) The sketch of the graph is shown here in Exercise 2 Figure 4.

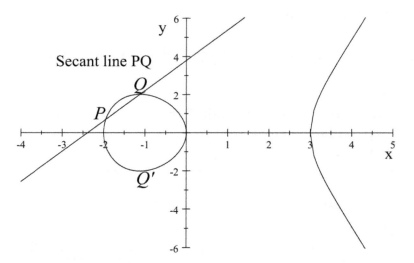

Exercise 2 Figure 4. A secant line PQ with slope m_{PQ}.

(b) To find the coordinates of the lowest and highest points, we start with the coordinates of P and Q.

$$P = (p, \sqrt{p(p+2)(p-3)}),$$
$$Q = (q, \sqrt{q(q+2)(q-3)}).$$

The slope of the secant line PQ is

$$m_{PQ} = \frac{\sqrt{p(p+2)(p-3)} - \sqrt{q(q+2)(q-3)}}{p-q}.$$

Notice that the numerator is the difference of two square roots; if we multiply the numerator and denominator by the sum of these same square roots, we get

$$m_{PQ} = \frac{p(p+2)(p-3) - q(q+2)(q-3)}{(p-q)\left(\sqrt{p(p+2)(p-3)} + \sqrt{q(q+2)(q-3)}\right)}.$$

After expanding the products and regrouping like terms in the numerator, then factoring and canceling the indeterminacy $(p-q)/(p-q)$ and we will get

$$m_{PQ} = \frac{(p^2 + pq + q^2) - (p+q) - 6)}{\sqrt{p(p+2)(p-3)} + \sqrt{q(q+2)(q-3)}}.$$

To get the slope of the tangent line at either Q or Q', we let $p \to q$, getting

$$m = \frac{3q^2 - 2q - 6}{2\sqrt{q(q+2)(q-3)}}.$$

We want to find the value of q that makes this slope $m = 0$. For m to be zero it must be true that $3q^2 - 2q - 6 = 0$. A fraction can be zero only if its numerator is zero. So $q = \frac{1-\sqrt{19}}{3} \approx -1.12$, as we found in Problem 2 of this homework exercise. The other root of the quadratic equation, $\frac{1+\sqrt{19}}{3}$ cannot be used since it would make $q(q+2)(q-3)$ negative, so we could not take its square root. Therefore the coordinates of the high point Q and the low point Q' are

$$Q = \left(\frac{1-\sqrt{19}}{3}, \sqrt{\frac{38\sqrt{19}-56}{27}} \right),$$

$$Q' = \left(\frac{1-\sqrt{19}}{3}, -\sqrt{\frac{38\sqrt{19}-56}{27}} \right).$$

Approximations to these coordinates are $Q = (-1.12, 2.015)$ and $Q' = (-1.12, -2.015)$. See Exercise 2 Figure 4.

3. Asymptotes

Answers to Exercises 3

1. The given equation is

$$y = \frac{(x-1)(x-2)}{(x-3)}.$$

(a) The graph, G is shown in Exercise 3, Figure 1. Also, in this figure, the vertical asymptote $x = 3$ is shown.

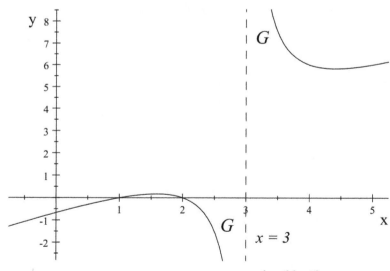

Exercise 3 Figure 1 Graph of $\frac{(x-1)(x-2)}{(x-3)}$.

(b) To find out what happens as $x \to \infty$, divide the numerator and

297

denominator by x, getting

$$y = \frac{(1 - \frac{1}{x})(x - 2)}{(1 - \frac{3}{x})}.$$

Now as $x \to \infty$, the factor $(1 - \frac{1}{x}) \to 1$, the factor $(1 - \frac{3}{x}) \to 1$ and the factor $(x - 2) \to \infty$, so $y \to \infty$ as $x \to \infty$.

(c) To find the coordinates of the highest point of G between $x = 1$ and $x = 2$, let's look at a close-up view of the graph G. See Exercise 3 Figure 2.

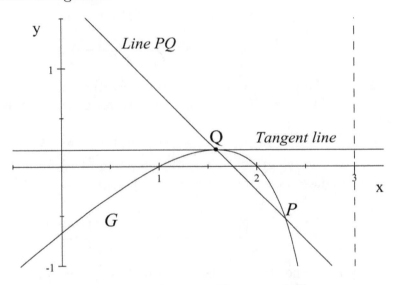

Exercise 3 Figure 2 Close up of G.

First, we set up the slope m_{PQ} of a line joining two points P and Q of G

$$m_{PQ} = \frac{\frac{(p-1)(p-2)}{(p-3)} - \frac{(q-1)(q-2)}{(q-3)}}{p - q},$$

which we re-write as,

$$m_{PQ} = \frac{(q - 3)(p - 1)(p - 2) - (p - 3)(q - 1)(q - 2)}{(p - q)(p - 3)(q - 3)}.$$

Multiply out the expressions in the numerator and simplify

$$m_{PQ} = \frac{pq(p - q) + 7(p - q) - 3(p + q)(p - q)}{(p - q)(p - 3)(q - 3)}.$$

Canceling the common factor $(p - q)$ from numerator and denominator, we get that the slope of the line PQ is

$$m_{PQ} = \frac{pq + 7 - 3(p + q)}{(p - 3)(q - 3)}.$$

If we let $P \to Q$ (so $p \to q$) we get that the secant line becomes a tangent line with slope m, where

$$m = \frac{q^2 - 6q + 7}{(q - 3)^2},$$

and if Q is the highest point the slope $m = 0$. In other words,

$$\frac{q^2 - 6q + 7}{(q - 3)^2} = 0.$$

But for this equation to be true, the numerator must be zero. That is, $q^2 - 6q + 7 = 0$. The two roots of this quadratic equation are

$$q_1 = 3 + \sqrt{2} \text{ and } q_2 = 3 - \sqrt{2}.$$

Where q_1 gives us the lowest point of G for $x > 3$, and q_2 the highest point of G for $x < 3$. (See Exercise 3, Figure 1.) We have

$$\frac{(q_2 - 1)(q_2 - 2)}{q_2 - 3} = \frac{(2 - \sqrt{2})(1 - \sqrt{2})}{-\sqrt{2}} = 3 - 2\sqrt{2}.$$

Thus the coordinates of the highest point Q are

$$(3 - \sqrt{2},\ 3 - 2\sqrt{2}).$$

2. Rapid sketch of the graph of the equation

$$y^2 = \frac{(x - 1)(x - 2)}{(x - 3)}.$$

This equation can be written

$$y = \pm\sqrt{\frac{(x - 1)(x - 2)}{(x - 3)}}.$$

The right hand side is $\pm$ the square root of the right hand side in Problem 1, so the graph is as shown in Exercise 3 Figure 3. We denote this graph as H.

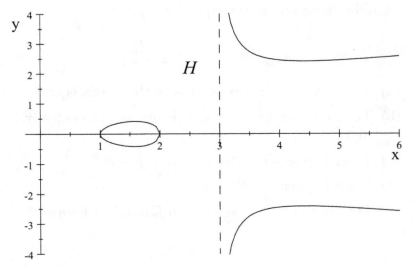

Exercise 3 Figure 3 Graph of $y^2 = \frac{(x-1)(x-2)}{x-3}$.

3. In Exercise 3, Figure 4 we sketch the graph of

$$y^2 = \frac{(x-1)(x-2)^2(x-3)}{(x-4)}.$$

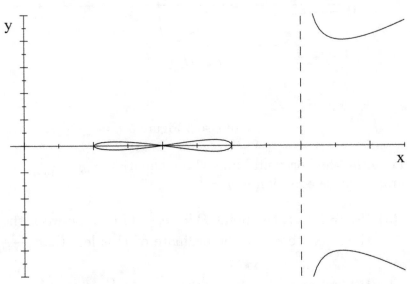

Exercise 3 Figure 4 Graph of $y^2 = \frac{(x-1)(x-2)^2(x-3)}{(x-4)}$.

4. Let K be the graph of the equation

$$y = \frac{x+1}{x^2+x+1}.$$

   (a) The only point where K crosses the x-axis is at $(-1,0)$.

   (b) The lowest and the highest points are, respectively $(-2, -\frac{1}{3})$ and $(0,1)$.

   (c) What happens to K as $x \to \infty$? $y \to 0$.

   (d) What happens to K as $x \to -\infty$? $y \to 0$.

   (e) The graph of K is sketched in Exercise 3 Figure. 5

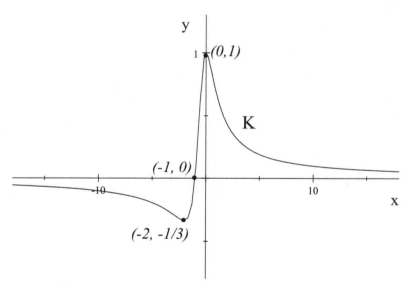

Exercise 3 Figure 5 $y = \frac{x+1}{x^2+x+1}$.

5. Let L be the horizontal line whose equation is $y = \frac{1}{10000}$, and G be the graph whose equation is $y = \frac{1}{x}$.

   (a) If $p = 10001$, the point $P = (p, \frac{1}{p})$ of G is between the line L and the x-axis because the ordinate of P is less than $\frac{1}{10000}$. That is, $0 < \frac{1}{p} < \frac{1}{10001} < \frac{1}{10000}$.

   (b) If t is any number larger than p then $\frac{1}{t} < \frac{1}{p} < \frac{1}{10000}$. Therefore all points of G to the right of P will be between L and the x-axis.

(c) If p is any positive number such that the point $P = (p, \frac{1}{p})$ is below L, then $\frac{1}{p} < \frac{1}{10000}$, that is $p > 10000$; there is always a smaller number q, for example $q = (10000 + p)/2$, such the point $Q = (q, \frac{1}{q})$ is between L and the x-axis. In other words

$$0 < \frac{2}{10000 + p} < \frac{1}{p} < \frac{1}{10000}.$$

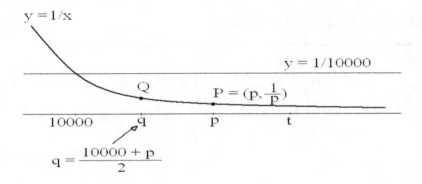

Exercise 3, Figure 6.

Hence, there is no left-most point of G between L and the x-axis. See Exercise 3, Figure 6.

4. Simple Graphs

Answers to Exercises 4

1. If $g(x) = 2x^4 + K$, where K is a constant, find the slope at $(a, g(a))$,

$$
\begin{aligned}
g'(a) &= \lim_{p \to a} \frac{g(p) - g(a)}{p - a} \\
&= \lim_{p \to a} \frac{2p^4 + K - (2a^4 + K)}{p - a}, \\
&= \lim_{p \to a} 2\frac{p^4 - a^4}{p - a}.
\end{aligned}
$$

Write $p - a$ as $(p - a)1$ and factor the difference between two fourth powers, thus,

$$
p^4 - a^4 = (p - a)(p^3 + p^2 a + p a^2 + a^3).
$$

So now we have

$$
g'(a) = \lim_{p \to a} 2\frac{(p - a)(p^3 + p^2 a + p a^2 + a^3)}{(p - a)1}.
$$

This means

$$
\begin{aligned}
g'(a) &= \lim_{p \to a} 2(p^3 + p^2 a + p a^2 + a^3), \\
&= 2(4a^3) = 8a^3.
\end{aligned}
$$

2. Given $f(x) = x^{1/2} = \sqrt{x}$,

(a) find $f'(x)$, for $x \neq 0$

$$f'(x) = \lim_{p \to x} \frac{f(p) - f(x)}{p - x},$$

$$= \lim_{p \to x} \frac{\sqrt{p} - \sqrt{x}}{p - x}.$$

We cannot let $p = x$, because, then we would get the indeterminate form $\frac{0}{0}$. In order to get rid of this meaningless expression, we must reduce the fraction to one in which letting $p \to x$ presents no problem. This can be achieved by multiplying the numerator and the denominator by $(\sqrt{p} + \sqrt{x})$ getting,

$$f'(x) = \lim_{p \to x} \frac{(\sqrt{p} - \sqrt{x})(\sqrt{p} + \sqrt{x})}{(p - x)(\sqrt{p} + \sqrt{x})}$$

$$= \lim_{p \to x} \frac{p - x}{(p - x)(\sqrt{p} + \sqrt{x})},$$

$$= \lim_{p \to x} \frac{1}{\sqrt{p} + \sqrt{x}},$$

$$f'(x) = \frac{1}{2\sqrt{x}} \text{ or } \frac{1}{2}x^{-1/2}.$$

(b) Does this graph have slope at the point $(0, 0)$? No, because for $f'(0)$ to exist, we would have.

$$f'(0) = \lim_{x \to 0} \frac{\sqrt{x} - \sqrt{0}}{x - 0}, \text{ but then}$$

$$f'(0) = \lim_{x \to 0} \frac{1}{\sqrt{x}} \text{ which does not exist.}$$

3. Here $h(x) = g(x) + f(x)$, where g and f are as in Problems 1 and 2, so

for $a \neq 0$,

$$
\begin{aligned}
h'(a) &= \lim_{p \to a} \frac{h(p) - h(a)}{p - a} \\
&= \lim_{p \to a} \frac{g(p) + f(p) - (g(a) + f(a))}{p - a}, \\
&= \lim_{p \to a} \left(\frac{g(p) - g(a)}{p - a} + \frac{f(p) - f(a)}{p - a} \right), \\
&= \lim_{p \to a} \frac{g(p) - g(a)}{p - a} + \lim_{p \to a} \frac{f(p) - f(a)}{p - a} = g'(a) + f'(a), \\
h'(a) &= 8a^3 + \frac{1}{2}a^{-1/2}.
\end{aligned}
$$

4. If $c(x) = 3h(x)$, then

$$
\begin{aligned}
c'(a) &= \lim_{p \to a} \frac{c(p) - c(a)}{p - a} \\
&= \lim_{p \to a} \frac{3h(p) - 3h(a)}{p - x}, \\
&= \lim_{p \to a} 3\frac{h(p) - h(a)}{p - a}, \\
&= 3h'(a), \\
&= 3(8a^3 + \frac{1}{2}a^{-1/2}), \\
c'(a) &= 24a^3 + \frac{3}{2}a^{-1/2}.
\end{aligned}
$$

5. For the equation, $y = r(x) = x(x - 1)(x + 1)$,

(a) The graph is shown in Exercise 4 Figure 1.

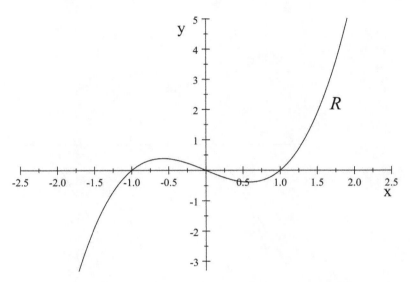

Exercise 4 Figure 1. Graph of $y = r(x) = x(x-1)(x+1)$.

(b) To get $r'(x)$, rewrite $r(x)$ as $x^3 - x$, then

$$
\begin{aligned}
r'(x) &= \lim_{t \to x} \frac{r(t) - r(x)}{t - x} \\
&= \lim_{t \to x} \frac{t^3 - t - (x^3 - x)}{t - x}, \\
&= \lim_{t \to x} \frac{t^3 - x^3 - (t - x)}{t - x}, \\
&= \lim_{t \to x} \frac{(t - x)(t^2 + tx + x^2) - (t - x)1}{t - x}, \\
&= 3x^2 - 1.
\end{aligned}
$$

(c) The slope of R is 5 when $3x^2 - 1 = 5$. Thus, $3x^2 = 6$, or $x = \pm\sqrt{2}$. Since $r(\sqrt{2}) = (\sqrt{2})^3 - \sqrt{2} = \sqrt{2}$, then one point where the slope is 5 is at $(\sqrt{2}, \sqrt{2})$. We call this point A in Exercise 4 Figure.2. We also have $r(-\sqrt{2}) = -\sqrt{2}$, so the other point where the graph R has slope 5, is $(-\sqrt{2}, -\sqrt{2})$ or point B in this same Figure.

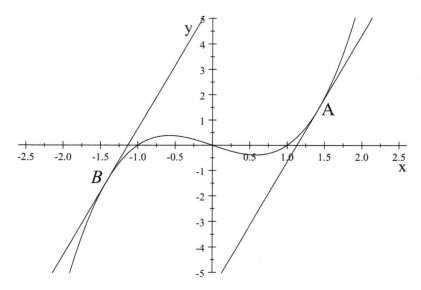

Exercise 4 Figure 2 Tangent lines to R with slope 5.

6. For the function $s(x) = x^3 + x^2 + x + 1$

(a) Find $s'(x)$

$$s'(x) = \lim_{p \to x} \frac{s(p) - s(x)}{p - x}$$

$$
\begin{aligned}
s'(x) &= \lim_{p \to x} \frac{p^3 + p^2 + p + 1 - (x^3 + x^2 + x + 1)}{p - x}, \\
&= \lim_{p \to x} \frac{p^3 - x^3 + p^2 - x^2 + p - x + 0}{p - x}.
\end{aligned}
$$

Now, factor

$$
\begin{aligned}
p^3 - x^3 &= (p - x)(p^2 + px + x^2) \\
p^2 - x^2 &= (p - x)(p + x), \text{ and} \\
p - x &= (p - x)1.
\end{aligned}
$$

So

$$
\begin{aligned}
s'(x) &= \lim_{p \to x} \frac{(p - x)(p^2 + px + x^2) + (p - x)((p + x) + (p - x)1}{p - x} \\
&\quad \lim_{p \to x} \frac{(p^2 + px + x^2) + (p + x) + 1}{1}, \\
&= 3x^2 + 2x + 1.
\end{aligned}
$$

(b) Can $s'(x)$ ever be zero? No, because $s'(x) = 3x^2 + 2x + 1$ and the equation $3x^2 + 2x + 1 = 0$ has no real root.

7. Given $u(t) = -\frac{1}{2}at^2 + v_0t + h$, where v_0 is any constant, and a and h are positive constants find where $u(t) = 0$. That is, if

$$-\frac{1}{2}at^2 + v_0t + h = 0,$$

find t by the quadratic formula

$$
\begin{aligned}
t &= \frac{-v_0 \pm \sqrt{v_0^2 - 4(-\frac{1}{2}a)(h)}}{2(-\frac{1}{2}a)}, \\
&= \frac{-v_0 \pm \sqrt{v_0^2 + 2ah}}{-a}.
\end{aligned}
$$

Denote these roots by t_1 and t_2

$$
\begin{aligned}
t_1 &= \frac{v_0 - \sqrt{v_0^2 + 2ah}}{a}, \\
t_2 &= \frac{v_0 + \sqrt{v_0^2 + 2ah}}{a}.
\end{aligned}
$$

(a) Since $\sqrt{v_0^2 + 2ah} > v_0$, then $t_1 < 0$ and $t_2 > 0$. Therefore, t_2 is the positive root.

(b) Now, for any t, we find $u'(t)$

$$
\begin{aligned}
u'(t) &= \lim_{p \to t} \frac{-\frac{1}{2}ap^2 + v_0p + h - (-\frac{1}{2}at^2 + v_0t + h)}{p - t}, \\
&= -at + v_0.
\end{aligned}
$$

So, for t_2,

$$
\begin{aligned}
u'(t_2) &= -at_2 + v_0, \\
&= -\sqrt{v_0^2 + 2ah}.
\end{aligned}
$$

5. Derivatives

Answers to Exercises 5

1. The natural domain for the function $f(x) = \frac{-1}{(x-2)(x-3)}$ includes all real numbers except $x = 2$ and $x = 3$. The domain designated in this problem, however, is merely the segment $(2, 3)$; so, we restrict the graph as shown in Exercise 5, Figure 1.

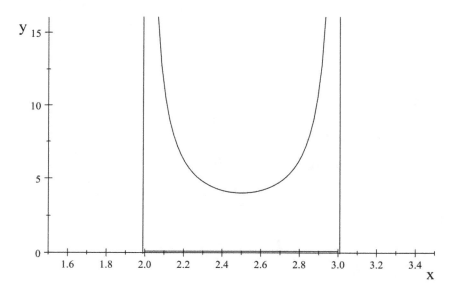

Exercise 5 Figure 1.

2. We can write $p - a$ as the difference of two cubes as follows: $p - a = (p^{1/3})^3 - (a^{1/3})^3$.

3. For $f(x) = x^{1/3}$,

309

(a) The natural domain is the set of all real numbers, since every real number has a real cube-root.

(b) The ordinate of P is $f(p) = p^{1/3}$.

(c) The ordinate of A is $f(a) = a^{1/3}$. The line PA has slope $(p^{1/3} - a^{1/3})/(p - a)$.

(d) We can factor the denominator $p - a$ as the difference of two cubes,

$$
\begin{aligned}
p - a &= (p^{1/3})^3 - (a^{1/3})^3, \\
&= (p^{1/3} - a^{1/3})(p^{2/3} + p^{1/3}a^{1/3} + a^{2/3}).
\end{aligned}
$$

The slope of PA is

$$
\begin{aligned}
\frac{p^{1/3} - a^{1/3}}{p - a} &= \frac{p^{1/3} - a^{1/3}}{(p^{1/3} - a^{1/3})(p^{2/3} + p^{1/3}a^{1/3} + a^{2/3})}, \\
&= \frac{1}{(p^{2/3} + p^{1/3}a^{1/3} + a^{2/3})}.
\end{aligned}
$$

After canceling the factor $p^{1/3} - a^{1/3}$ from the numerator and denominator, we can let $p \to a$.

$$
\begin{aligned}
f'(a) &= \frac{1}{3a^{2/3}}, \text{ or} \\
f'(a) &= \frac{1}{3}a^{-2/3}.
\end{aligned}
$$

4. By definition of derivative (Equation 5.1)

$$
\begin{aligned}
f'(x) &= \lim_{t \to x} \frac{f(t) - f(x)}{t - x} \\
&= \lim_{t \to x} \frac{t^n - x^n}{t - x}.
\end{aligned}
$$

The expression $t^n - x^n$ is the difference of *two nth-powers*, and can be factored as follows:

$$
t^n - x^n = (t - x)(t^{n-1} + t^{n-2}x + t^{n-3}x^2 + \ldots + x^{n-1}).
$$

This means we can write

$$
\frac{t^n - x^n}{t - x} = \frac{(t - x)(t^{n-1} + t^{n-2}x + t^{n-3}x^2 + \ldots + x^{n-1})}{t - x}.
$$

After canceling the factor $t - x$ from numerator and denominator, we can let $t \to x$ and we get

$$f'(x) = nx^{n-1},$$

for any positive integer n.

5. By definition of derivative

$$
\begin{aligned}
g'(x) &= \lim_{t \to x} \frac{g(t) - g(x)}{t - x} \\
&= \lim_{t \to x} \frac{t^{1/n} - x^{1/n}}{t - x}.
\end{aligned}
$$

Now let us rewrite the denominator $t - x$ as the difference of two nth-powers

$$
\begin{aligned}
t - x &= (t^{1/n})^n - (x^{1/n})^n \\
&= (t^{1/n} - x^{1/n})\left((t^{1/n})^{n-1} + (t^{1/n})^{n-2}(x^{1/n}) + \ldots + (x^{1/n})^{n-1}\right), \\
&= (t^{1/n} - x^{1/n})(t^{1-1/n} + t^{1-2/n}x^{1/n} + t^{1-3/n}x^{2/n} + \ldots + x^{1-1/n}).
\end{aligned}
$$

Replacing the denominator by this version of $t - x$, we get

$$g'(x) = \lim_{t \to x} \frac{t^{1/n} - x^{1/n}}{(t^{1/n} - x^{1/n})(t^{1-1/n} + t^{1-2/n}x^{1/n} + t^{1-3/n}x^{2/n} + \ldots + x^{1-1/n})}.$$

Now we can cancel the $t^{1/n} - x^{1/n}$ out of both the numerator and denominator, getting

$$
\begin{aligned}
g'(x) &= \lim_{t \to x} \frac{1}{(t^{1-1/n} + t^{1-2/n}x^{1/n} + t^{1-3/n}x^{2/n} + \ldots + x^{1-1/n})}, \\
&= \frac{1}{nx^{1-1/n}},
\end{aligned}
$$

which is usually written as

$$g'(x) = \frac{1}{n}x^{1/n-1}.$$

6. For $g(x) = x^{-2}$

(a) The line AP, joining the two points $A = (a, g(a))$ and $P = (p, g(p))$, has slope

$$
\begin{aligned}
m_{AP} &= \frac{g(a) - g(p)}{a - p} \\
&= \frac{a^{-2} - p^{-2}}{a - p}, \\
&= \frac{\frac{1}{a^2} - \frac{1}{p^2}}{a - p} = \frac{\frac{p^2 - a^2}{a^2 p^2}}{a - p}, \\
&= \frac{p^2 - a^2}{a^2 p^2 (a - p)}, \\
&= \frac{(p - a)(p + a)}{a^2 p^2 (a - p)}.
\end{aligned}
$$

(b) This simplifies to $-(p + a)/(a^2 p^2)$ now we can let $p \to a$, getting $-2a/a^4$, or $-2a^{-3}$ for the slope of G at the point A.

7. If $h(x) = x^5 + x^4 + x^3 + x^2 + x^1 + 1 + x^{-1} + x^{-2}$, then from the fact that the derivative of the sum of two functions is the sum of their derivatives and the derivative of x^n is nx^{n-1}, we can infer that

$$
h'(x) = 5x^4 + 4x^3 + 3x^2 + 2x + 1 + 0 - x^{-2} - 2x^{-3}.
$$

8. If $f(x) = x^{100}$, then

(a) $f'(x) = 100x^{99}$, if $g(x) = 5f(x)$, then $g(x) = 5x^{100}$, so

$$
\begin{aligned}
g'(x) &= \lim_{t \to x} \frac{5t^{100} - 5x^{100}}{t - x} \\
&= \lim_{t \to x} 5 \frac{t^{100} - x^{100}}{t - x}, \\
&= 5 \times \lim_{t \to x} \frac{t^{100} - x^{100}}{t - x}, \\
&= 5 \times 100x^{99}, \\
g'(x) &= 500x^{99}.
\end{aligned}
$$

Also, if $h(x) = Cf(x)$, then

$$
\begin{aligned}
h'(x) &= \lim_{t \to x} \frac{Cf(t) - Cf(x)}{t - x} \\
&= \lim_{t \to x} C \frac{f(t) - f(x)}{t - x} = Cf'(x), \\
&= 100Cx^{99}.
\end{aligned}
$$

(b) Let $p(x)$ and $q(x)$ be two functions with derivatives at each point on the same domain D, and let $r(x) = p(x) - q(x)$, then by definition of derivative

$$
\begin{aligned}
r'(x) &= \lim_{t \to x} \frac{r(t) - r(x)}{t - x} \\
&= \lim_{t \to x} \frac{p(t) - q(t) - (p(x) - q(x))}{t - x}, \\
&= \lim_{t \to x} \frac{p(t) - p(x) - (q(t) - q(x))}{t - x}, \\
&= \lim_{t \to x} \frac{p(t) - p(x)}{t - x} - \lim_{t \to x} \frac{q(t) - q(x)}{t - x}, \\
&= p'(x) - q'(x).
\end{aligned}
$$

9. No, because for example, if $f(x) = x^3$, $g(x) = x^4$, and $h(x) = f(x)g(x)$, then $h(x) = x^7$ and $h'(x) = 7x^6$. But $f'(x) = 3x^2$, and $g'(x) = 4x^3$. So, $f'(x)g'(x) = 12x^5$, which is not $h'(x)$.

6. Derivative Formulas

Answers to Exercises 6

1. Find $(x^{m/n})'$. There are several ways to solve this problem. We will show two of them.

Method 1.

$$
\begin{aligned}
(x^{m/n})' &= \lim_{t \to x} \frac{t^{m/n} - x^{m/n}}{t - x} \\
&= \lim_{t \to x} \frac{t^{m/n} - x^{m/n}}{t^{n/n} - x^{n/n}}, \\
&= \lim_{t \to x} \frac{(t^{1/n} - x^{1/n})((t^{1/n})^{m-1} + (t^{1/n})^{m-2}x^{1/n} + \ldots + (t^{1/n})^{m-1})}{(t^{1/n} - x^{1/n})((t^{1/n})^{n-1} + (t^{1/n})^{n-2}x^{1/n} + \ldots + (t^{1/n})^{n-1})}, \\
&= \lim_{t \to x} \frac{((t^{1/n})^{m-1} + (t^{1/n})^{m-2}x^{1/n} + \ldots + (t^{1/n})^{m-1})}{((t^{1/n})^{n-1} + (t^{1/n})^{n-2}x^{1/n} + \ldots + (t^{1/n})^{n-1})}, \\
&= \frac{x^{m/n-1/n} + x^{m/n-1/n} + \ldots + x^{m/n-1/n}}{x^{1-1/n} + x^{1-1/n} + \ldots + x^{1-1/n}}, \\
&= \frac{m x^{m/n-1/n}}{n x^{1-1/n}} = \frac{m}{n} x^{m/n-1}.
\end{aligned}
\tag{4}
$$

Method 2.

$$
\begin{aligned}
(x^{m/n})' &= \lim_{t \to x} \frac{t^{m/n} - x^{m/n}}{t - x} \\
&= \lim_{t \to x} \left(\frac{t^{m/n} - x^{m/n}}{t^{1/n} - x^{1/n}} \right) \left(\frac{t^{1/n} - x^{1/n}}{t - x} \right), \\
&= \lim_{t \to x} \frac{t^{m/n} - x^{m/n}}{t^{1/n} - x^{1/n}} \lim_{t \to x} \frac{t^{1/n} - x^{1/n}}{t - x}.
\end{aligned}
$$

Now let $t^{1/n} = r$ and $x^{1/n} = s$ then $t^{m/n} = r^m$, and $x^{m/n} = s^m$. As t approaches x, r will approach s. Therefore,

$$\lim_{t \to x} \frac{t^{m/n} - x^{m/n}}{t^{1/n} - x^{1/n}} = \lim_{r \to s} \frac{r^m - s^m}{r - s}.$$

So, we can write

$$(x^{m/n})' = \lim_{r \to s} \frac{r^m - s^m}{r - s} \lim_{t \to x} \frac{t^{1/n} - x^{1/n}}{t - x},$$

or

$$(x^{m/n})' = ms^{m-1} \times \frac{1}{n} x^{(1/n)-1}$$

$$(x^{m/n})' = m(x^{(m/n)-(1/n)}) \times \frac{1}{n} x^{(1/n)-1},$$

$$= \frac{m}{n} x^{m/n-1}.$$

2. From the definition of derivative

$$(f(x)g(x))' = \lim_{t \to x} \frac{f(t)g(t) - f(x)g(x)}{t - x}.$$

Now use the "old mathematicians trick" of changing what you've got to what you'd rather have. Here we subtract and add $f(t)g(x)$ to the numerator.

$$(f(x)g(x))' = \lim_{t \to x} \frac{f(t)g(t) - f(t)g(x) + f(t)g(x) - f(x)g(x)}{t - x}$$

$$= \lim_{t \to x} \left[\frac{f(t)g(t) - f(t)g(x)}{t - x} + \frac{f(t)g(x) - f(x)g(x)}{t - x} \right],$$

$$= \lim_{t \to x} f(t) \frac{g(t) - g(x)}{t - x} + \lim_{t \to x} g(x) \frac{f(t) - f(x)}{t - x}.$$

In the first term, the limit of $\frac{f(t)-f(x)}{t-x}$ as $t \to x$ is $f'(x)$, and in the second, the limit of $\frac{g(t)-g(x)}{t-x}$ as $t \to x$ is $g'(x)$. So,

$$(f(x)g(x))' = f(x)g'(x) + g(x)f'(x).$$

The verbal expression of this is, "The derivative of the product of two functions is the first times the derivative of the second plus the second times the derivative of the first".

3. The derivative of the quotient $\frac{f(x)}{g(x)}$, by definition, is as follows:

$$\left(\frac{f(x)}{g(x)}\right)' = \lim_{t \to x} \frac{\frac{f(t)}{g(t)} - \frac{f(x)}{g(x)}}{t - x},$$

$$= \lim_{t \to x} \frac{f(t)g(x) - g(t)f(x)}{(t - x)g(t)g(x)}.$$

Now we will use "the old mathematicians trick" of adding zero; this time by subtracting and adding $f(x)g(x)$ to the numerator.

$$\left(\frac{f(x)}{g(x)}\right)' = \lim_{t \to x} \frac{f(t)g(x) - f(x)g(x) + f(x)g(x) - g(t)f(x)}{(t - x)g(t)g(x)},$$

$$= \lim_{t \to x} \frac{g(x)\,(f(t) - f(x))}{(t - x)g(t)g(x)} - \lim_{t \to x} \frac{f(x)\,(g(t) - g(x))}{(t - x)g(t)g(x)}.$$

But, now

$$\lim_{t \to x} \frac{g(x)\,(f(t) - f(x))}{(t - x)g(t)g(x)} = \frac{g(x)}{(g(x))^2}f'(x),$$

and

$$\lim_{t \to x} \frac{f(x)\,(g(t) - g(x))}{(t - x)g(t)g(x)} = \frac{f(x)}{(g(x))^2}g'(x),$$

therefore

$$\left(\frac{f(x)}{g(x)}\right)' = \frac{g(x)f'(x) - f(x)g'(x)}{(g(x))^2}.$$

4. The verbal expression is

"The derivative of the quotient of two functions is the denominator times the derivative of the numerator, minus the numerator times the derivative of the denominator, divided by the denominator squared".

5. To find the derivative of $\frac{1}{f(x)}$, we write the denominator $f(x)$ times 0 (the derivative of the numerator) minus 1 (the numerator) times the derivative of the denominator $f'(x)$ divided by the denominator squared, $f(x)^2$. In other words,

$$\frac{f(x) \times 0 - 1 \times f'(x)}{f(x)^2} = \frac{-f'(x)}{f(x)^2}.$$

6. For $y = \frac{x-1}{(x-2)(x-3)}$,

(a) the graph is shown in Exercise 6 Figure 1.

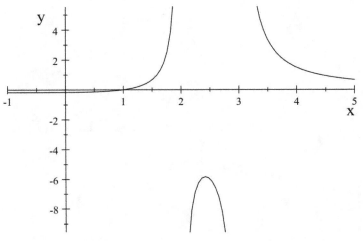

Exercise 6 Figure 1 Graph of $y = \frac{x-1}{(x-2)(x-3)}$.

(b) The slope of this graph at any point $(x, y(x))$ is

$$y'(x) = \frac{-x^2 + 2x + 1}{(x-2)^2(x-3)^2}.$$

7. If $h(x) = \frac{1}{x}$, then

(a) The graph of $y = h(x)$ is shown in Exercise 6, Figure 2.

(b) $h'(x) = \frac{-1}{x^2}$. The graph of $y = h'(x)$ is also shown in this same figure.

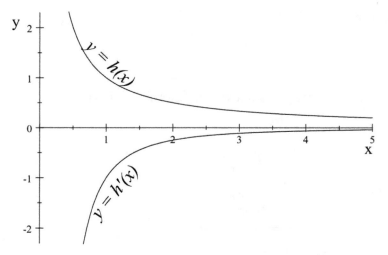

Exercise 6, Figure 2 Graphs of $h(x)$ and $h'(x)$.

7. The Limit of $\sin(x)/x$ as $x \to 0$

Answers to Exercises 7

1. To find $\lim_{x \to 0} \frac{\sin(x)}{\sqrt{x}}$, first multiply the numerator and denominator by $\sqrt{x}$,

$$\lim_{x \to 0} \frac{\sin(x)}{\sqrt{x}} = \lim_{x \to 0} \frac{\sqrt{x}\sin(x)}{x},$$

then

$$\begin{aligned}
\lim_{x \to 0} \frac{\sin(x)}{\sqrt{x}} &= \lim_{x \to 0} \frac{\sqrt{x}}{1} \times \lim_{x \to 0} \frac{\sin(x)}{x}, \\
&= 0 \times 1 = 0.
\end{aligned}$$

2. To find $\lim_{x \to 0} \frac{\sin(\sqrt{x})}{x}$, first rewrite the denominator as $\sqrt{x} \times \sqrt{x}$,

$$\lim_{x \to 0} \frac{\sin(\sqrt{x})}{x} = \lim_{x \to 0} \frac{\sin(\sqrt{x})}{\sqrt{x} \times \sqrt{x}},$$

then

$$\begin{aligned}
\lim_{x \to 0} \frac{\sin(\sqrt{x})}{x} &= \lim_{x \to 0} \frac{1}{\sqrt{x}} \times \lim_{x \to 0} \frac{\sin(\sqrt{x})}{\sqrt{x}}, \\
&= \infty \times 1 = \infty.
\end{aligned}$$

3. If $f(x) = \cos(x)$, we find $f'(x)$ by the definition of derivative,

$$f'(x) = \lim_{t \to x} \frac{\cos(x) - \cos(t)}{x - t},$$

321

let $x = p + q$ and $t = p - q$, then $x - t = 2q$, and $q \to 0$ as $t \to x$, so

$$
\begin{aligned}
f'(x) &= \lim_{q \to 0} \frac{\cos(p + q) - \cos(p - q)}{2q} \\
&= \lim_{q \to 0} \frac{-2\sin(p)\sin(q)}{2q}, \\
&= \lim_{q \to 0} \frac{-\sin(p)}{1} \times \lim_{q \to 0} \frac{\sin(q)}{q}, \\
&= \lim_{q \to 0} -\sin(p).
\end{aligned}
$$

But what does p approach as $q \to 0$? Since $p = x - q$, then $p \to x$, as $q \to 0$. So the derivative of $\cos(x)$ is $-\sin(x)$.

4. Since

$$
\tan(x) = \frac{\sin(x)}{\cos(x)},
$$

then by the formula for the derivative of a quotient of two functions, we can write

$$
(\tan(x))' = \frac{\cos(x) \times (\sin(x))' - \sin(x) \times (\cos(x))'}{(\cos(x))^2},
$$

$$
= \frac{\cos(x) \times \cos(x) + \sin(x) \times \sin(x)}{\cos^2(x)},
$$

$$
= \frac{1}{\cos^2(x)}.
$$

So, $(\tan(x))' = \sec^2(x)$.

5. Since

$$
\csc(x) = \frac{1}{\sin(x)},
$$

then

$$
(\csc(x))' = \frac{-(\sin(x))'}{\sin^2(x)} = \frac{-\cos(x)}{\sin^2(x)} = -\cot(x)\csc(x).
$$

6. Since

$$\sec(x) = \frac{1}{\cos(x)},$$

then

$$(\sec(x))' = \frac{-(\cos(x))'}{\cos^2(x)} = \frac{\sin(x)}{\cos^2(x)} = \tan(x)\sec(x).$$

7. Since

$$\cot(x) = \frac{\cos(x)}{\sin(x)},$$

then

$$\begin{aligned} (\cot(x))' &= \frac{\sin(x) \times (\cos(x))' - \cos(x) \times (\sin(x))'}{\sin^2(x)}, \\ &= \frac{-\sin^2(x) - \cos^2(x)}{\sin^2(x)} = -\csc^2(x). \end{aligned}$$

8. Here, we don't have a formula to relate $\sin(x^3)$ to other trig functions. In a case such as this we can always go back to the definition of derivative.

$$(\sin(x^3))' = \lim_{t \to x} \frac{\sin(x^3) - \sin(t^3)}{x - t}.$$

Multiply and divide by $(x^3 - t^3)$

$$\begin{aligned} (\sin(x^3))' &= \lim_{t \to x} \frac{\sin(x^3) - \sin(t^3)}{x^3 - t^3} \times \frac{x^3 - t^3}{x - t} \\ &= \lim_{t \to x} \frac{\sin(x^3) - \sin(t^3)}{x^3 - t^3} \times \lim_{t \to x} \frac{x^3 - t^3}{x - t}, \\ &= \cos(x^3) \times 3x^2. \end{aligned}$$

9. To find $(\sin(\sqrt{x}))'$, we can use our "fall-back" method, as we did in the previous problem. By the definition of derivative,

$$(\sin(\sqrt{x}))' = \lim_{t \to x} \frac{\sin(\sqrt{x}) - \sin(\sqrt{t})}{x - t},$$

multiply and divide by $(\sqrt{x} - \sqrt{t})$

$$
\begin{aligned}
(\sin(\sqrt{x}))' &= \lim_{t \to x} \frac{\sin(\sqrt{x}) - \sin(\sqrt{t})}{\sqrt{x} - \sqrt{t}} \times \frac{\sqrt{x} - \sqrt{t}}{x - t} \\
&= \lim_{t \to x} \frac{\sin(\sqrt{x}) - \sin(\sqrt{t})}{\sqrt{x} - \sqrt{t}} \times \lim_{t \to x} \frac{\sqrt{x} - \sqrt{t}}{x - t}, \\
&= \cos(\sqrt{x}) \times \frac{1}{2} x^{-1/2}.
\end{aligned}
$$

8. The Chain Rule

Answers to Exercises 8

1. If $y = (f(x))^{100}$, then $y'(x) = 100(f(x))^{99} f'(x)$.

2. If $y(x) = f(g(h(x)))$, then by the chain rule
$$\frac{df(g(h(x)))}{dx} = \frac{df(g(h(x)))}{dg(h(x))} \frac{dg(h(x))}{dh(x)} \frac{dh(x)}{dx}.$$

3. If $y(x) = \tan(\cos(\sin(x)))$, then
$$\frac{dy(x)}{dx} = \sec^2(\cos(\sin(x))) \times (-\sin(\sin(x))) \times \cos(x).$$

4. If $y = \cos(x^3 + 1)$, then
$$\begin{aligned} y'(x) &= -\sin(x^3 + 1) \times 3x^2, \text{ or} \\ y'(x) &= -3x^2 \sin(x^3 + 1). \end{aligned}$$

5. If $y = \cos^3(ax)$, where a is a constant, then
$$y'(x) = -3a \cos^2(ax) \sin(ax).$$

6. If $y = (x^3 - 7x^2 + 10 + \sin(x))^{13}$, then
$$y'(x) = 13(x^3 - 7x^2 + 10 + \sin(x))^{12} \times (3x^2 - 14x + \cos(x)).$$

7. Differentiate $(x^7 - \frac{y(x)}{x})^{5/8}$
$$\frac{5}{8}(x^7 - \frac{y(x)}{x})^{-3/8} \times (7x^6 - \frac{xy'(x) - y(x)}{x^2}).$$

8. First, rewrite the equation $\sqrt{\cos(x) + \sin(x)} - y(x) = 0$ as $y(x) = \sqrt{\cos(x) + \sin(x)}$, then

$$\frac{dy(x)}{dx} = \frac{-\sin(x) + \cos(x)}{2\sqrt{\cos(x) + \sin(x)}}.$$

9. If $y = \sqrt{\frac{\sin(x)}{\cos(x)}}$, then $y = \sqrt{\tan(x)}$, so

$$\frac{dy}{dx} = \frac{1}{2}(\tan(x))^{-1/2}\sec^2(x).$$

10. If $y = \sqrt{\sin(1/\cos(x^2))}$ then

$$\frac{dy}{dx} = (\sin(1/\cos(x^2)))^{-1/2}\cos(1/\cos(x^2))\frac{\sin(x^2)}{\cos^2(x^2)}x.$$

11. We are given that $x(t) = \sin(t)$, $y(x(t)) = u(t)$, $y'(x) = \sqrt{1 - x^2}$, find $u'(t)$

$$\frac{dx}{dt} = \cos(t),$$

$$\frac{dy}{dx}\frac{dx}{dt} = \frac{du}{dt}, \text{ or}$$

$$\frac{dy}{dx}\cos(t) = \frac{du}{dt}.$$

But we are also given that $\frac{dy}{dx} = \sqrt{1 - x^2}$, which is $\sqrt{1 - (\sin(t))^2}$, or $\cos(t)$. Hence,

$$\frac{du}{dt} = \cos^2(t).$$

12. By the chain rule, dy/dx is

$$\frac{dg\,[f(x)h(x) + k(u(x))]}{d\,[f(x)h(x) + k(u(x))]}\frac{d\,[f(x)h(x) + k(u(x))]}{dx}$$

$$\frac{dg\,[f(x)h(x) + k(u(x))]}{d\,[f(x)h(x) + k(u(x))]}\,(f'(x)h(x) + h'(x)f(x) + k'(u)u'(x)).$$

9. Implicit Differentiation of an Equation

Answers to Exercises 9

1. We are given that $xy \sin(y) = y^2 + x + 1$. Differentiating both sides with respect to x, we get

$$\frac{d(xy \sin(y))}{dx} = \frac{d(y^2 + x + 1)}{dx},$$
$$xy \cos(y)y'(x) + xy'(x) \sin(y) + y \sin(y) = 2yy'(x) + 1.$$

Now, solving for $y'(x)$, we get

$$y'(x) = \frac{1 - y \sin(y)}{(xy \cos(y) + x \sin(y) - 2y)}.$$

2. If $y^2 = \sin(\frac{1}{\cos^2(x)})$, then $y^2 = \sin(\sec^2(x))$

$$2yy' = \cos(\sec^2(x))2 \sec(x) \sec(x) \tan(x)$$
$$\frac{dy}{dx} = \frac{\sec^2(x) \tan(x) \cos(\sec^2(x))}{y},$$
$$= \frac{\sec^2(x) \tan(x) \cos(\sec^2(x))}{\pm\sqrt{\sin(\sec^2(x))}}.$$

3. Rewrite the equation $x = \frac{y^2 + 3y + 1}{\sec(y)}$ as

$$x \sec(y) = y^2 + 3y + 1.$$

then differentiating we get

$$x \sec(y) \tan(y) y' + \sec(y) = 2yy' + 3y'$$

$$y'(x) = \frac{\sec(y)}{2y + 3 - x \sec(y) \tan(y)},$$

$$\frac{dy}{dx} = \frac{\sec(y)}{2y + 3 - x \sec(y) \tan(y)}.$$

4. The equation $\alpha(x) = \cos^{-1}(x)$ means that $\cos(\alpha(x)) = x$, and since $\sin^2(\alpha) + \cos^2(\alpha) = 1$, then

$$\sin(\alpha(x)) = \pm\sqrt{1 - \cos^2(\alpha(x))} = \pm\sqrt{1 - x^2},$$

so

$$\frac{d\sin(\alpha(x))}{dx} = \pm\frac{1}{2}(1 - x^2)^{-1/2}(-2x),$$

$$= \mp x(1 - x^2)^{-1/2}.$$

5. If $y = \sqrt{\sin(\sec(x^2))}$, then

$$\frac{dy(x)}{dx} = \frac{1}{2}(\sin(\sec(x^2)))^{-1/2} \times \cos(\sec(x^2)) \times \sec(x^2) \tan(x^2) \times 2x,$$

$$= \frac{x \cos(\sec(x^2)) \sec(x^2) \tan(x^2)}{\sqrt{\sin(\sec(x^2))}}.$$

6. If $x^2 + 3xy + y^2 = \tan(y(x))$, then by implicit differentiation

$$2x + 3x\frac{dy}{dx} + 3y + 2y\frac{dy}{dx} = \sec^2(y(x))\frac{dy}{dx},$$

which simplifies to

$$\frac{dy}{dx} = \frac{-2x - 3y}{3x + 2y - \sec^2(y)}.$$

7. If $\sqrt{x^2 + y^2} = 9$, then

$$\frac{1}{2}(x^2 + y^2)^{-1/2}(2x + 2y\frac{dy}{dx}) = 0,$$

$$\frac{(x + y\frac{dy}{dx})}{\sqrt{x^2 + y^2}} = 0.$$

Since a fraction is zero only if the numerator is zero, then

$$x + y\frac{dy}{dx} = 0, \text{ or}$$
$$\frac{dy}{dx} = \frac{-x}{y}.$$

10. Inverse Functions

Answers to Exercises 10

1. If $u(x) = \cos^{-1}(x)$, find $u'(x)$. Rewrite the equation as $\cos(u(x)) = x$, now by implicit differentiation

$$- \sin(u(x)) \frac{du}{dx} = 1,$$

so

$$\frac{du}{dx} = - \frac{1}{\sin(u(x))}.$$

But, since $\cos(u(x)) = x$, then $\sin(u(x)) = \pm\sqrt{1 - x^2}$, therefore

$$\frac{du}{dx} = \frac{\pm 1}{\sqrt{1 - x^2}}.$$

2. We are given $z(t) = \sec^{-1}(t)$. Rewrite as $t = \sec(z(t))$ and differentiate with respect to t

$$1 = \sec(z(t)) \tan(z(t)) \frac{dz(t)}{dt},$$

so

$$\frac{dz(t)}{dt} = \frac{1}{\sec(z(t)) \tan(z(t))}.$$

But, from $\sec(z(t)) = t$, we conclude that $\tan(z(t)) = \pm\sqrt{t^2 - 1}$, therefore

$$\frac{dz(t)}{dt} = \frac{\pm 1}{t\sqrt{t^2 - 1}}.$$

331

3. If $y(x) = \sin^{-1}(\sqrt{x})$, then

$$y'(x) = \frac{\pm 1}{\sqrt{1-x}} \times \frac{1}{2}x^{-1/2},$$

or (one of many simplifications)

$$y'(x) = \frac{1}{2\sqrt{x - x^2}}.$$

4. If $y(x) = \tan^{-1}(f(x))$, then

$$y'(x) = \frac{f'(x)}{1 + f^2(x)}.$$

5. If $y(x) = (\sin^{-1}(x))^3$, then

$$\frac{dy(x)}{dx} = \pm\frac{3(\sin^{-1}(x))^2}{\sqrt{1-x^2}}.$$

6. Since f and g are inverses of each other then

$$f(g(x)) = x,$$

therefore, by the chain rule

$$\frac{df}{dg}\frac{dg}{dx} = 1, \text{ so}$$

$$\frac{df}{dg} = \frac{1}{\frac{dg}{dx}}.$$

7. Problems (a), (b), and (c) are stated and followed by their answers.

(a) If $a \neq 0$ and $u(x) = \tan^{-1}(\frac{x}{a})$, find $u'(x)$. Simplify as much as possible. To solve this, we first rewrite the equation as

$$u(x) = \tan^{-1}(\frac{x}{a}),$$

$$\tan(u(x)) = \frac{x}{a},$$

then by implicit differentiation

$$\sec^2(u(x))u'(x) = \frac{1}{a}. \tag{I}$$

Here we notice that the equation $\tan(u(x)) = x/a$ can be pictured in a right triangle, so that

$$\sec(u(x)) = \frac{h}{a} = \frac{\sqrt{a^2 + x^2}}{a}. \tag{II}$$

See Exercise 10, Figure 1

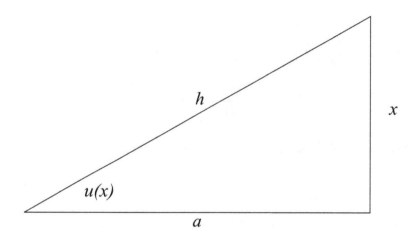

Exercise 10, Figure 1 $h = \sqrt{a^2 + x^2}$.

From Equations (I) and (II)

$$\frac{(a^2 + x^2)}{a^2}u'(x) = \frac{1}{a},$$

so

$$u'(x) = \frac{a}{a^2 + x^2}. \tag{III}$$

(b) Find a function $u(t)$ such that

$$u'(t) = \frac{k}{k^2 + (t + h)^2}. \tag{IV}$$

To solve this we notice that in Equation (IV), we've got k^2 and $(t+h)^2$, and we would rather have a^2 and x^2 as we have in Equation (III). Let's set $a = k$ and $x = t + h$

$$u(t) = \tan^{-1}(\frac{t+h}{k}). \tag{V}$$

Differentiating Equation (V) gives us Equation (IV).

(c) Find a function $y(x)$ such that

$$y'(x) = \frac{1}{x^2 + x + 1}.$$

We can solve this by finding two constants h and k such that $x^2 + x + 1 = k^2 + (x + h)^2$. The solution is

$$y(x) = \frac{4}{\sqrt{3}} \tan^{-1}\left(\frac{2x + 1}{\sqrt{3}}\right).$$

8. If $f(x) = g(h(x))$, then by the chain rule

$$\frac{df(x)}{dx} = \frac{dg(x)}{dh(x)}\frac{dh(x)}{dx},$$

which we could write as

$$\frac{df}{dx} = \frac{dg}{dh}\frac{dh}{dx}, \tag{VI}$$

with the understanding that $f = f(x)$, $g = g(h)$, and $h = h(x)$. Divide both sides by $\frac{dh}{dx}$, and we get

$$\frac{dg}{dh} = \frac{\frac{df}{dx}}{\frac{dh}{dx}}.$$

11. Second Derivative

Answers to Exercises 11

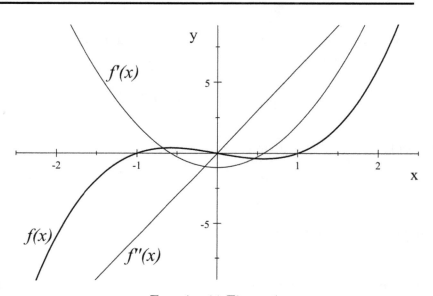

Exercise 11 Figure 1.

1. If $f(x) = (x - 1)x(x + 1) = x^3 - x$, then

   (a)
   $$\frac{df(x)}{dx} = 3x^2 - 1,$$
   $$\frac{d^2 f(x)}{dx} = 6x.$$

   (b) The graphs of $y_1 = f(x)$, $y_2 = f'(x)$ and $y_3 = f''(x)$ are shown in Exercise 11 Figure 1.

(c) There are two points of the graph where $f'(x) = 0$.

$$f'(x) = 0, \text{ when}$$
$$x_1 = -\frac{1}{\sqrt{3}} \text{ and } x_2 = +\frac{1}{\sqrt{3}},$$

so at x_1 and x_2,

$$f''(x_1) = -6/\sqrt{3} \text{ and } f''(x_2) = +6/\sqrt{3}.$$

2. If $y(x) = \sin(ax)$, then $y'(x) = a\cos(ax)$ and $y''(x) = -a^2\sin(ax)$.

3. If $y(x) = \cos(3x)$, then $y''(x) = -9\cos(3x)$, therefore $y''(x) + 9y(x) = 0$.

4. If $g(x) = \frac{1}{4}x^4 - 2x^3 + \frac{11}{2}x^2 - 6x + 10$, then $g'(x) = x^3 - 6x^2 + 11x - 6$.

   (a) $g'(x)$ factors into $(x-1)(x-2)(x-3)$. Therefore $g'(x) = 0$ implies that $(x-1)(x-2)(x-3) = 0$. So our three roots for $g'(x)$ are $x_1 = 1$, $x_2 = 2$, and $x_3 = 3$.

   (b) $g''(x) = 3x^2 - 12x + 11$, therefore $g''(1) = 2 > 0$, $g''(2) = -1 < 0$, and $g''(3) = 2 > 0$.

   (c) The graph of the equation $y = g(x)$ has a relative minimum at $x = 1$, a relative maximum at $x = 2$, and a relative minimum at $x = 3$.

5. Using the formula for the velocity $v(t)$ (in Km./hour) of the Shinkansen bullet train

$$v(t) = \begin{cases} 280t^{1/2} & \text{if } 0 \le t < 1 \\ 280 + 0.1(16 - (t-2)^4)^{1/4} & \text{if } 1 < t < 3 \\ 280(4-t)^{1/2} & \text{if } 3 < t \le 4 \end{cases}$$

   (a) We see that $v(0) = 0$ Km./hour, and $v(1/2) = 280 \times \sqrt{1/2} \approx 198$ Km./hour. For $t = 2.5$, we get $v(2.5) \approx 280.2$ Km./hour. 3 hours, 50 minutes translates to 3.833 hours, so $v(3.833) \approx 114.3$ Km./hour.

(b) The acceleration equation is obtained by differentiating the given velocity equation

$$v'(t) = a(t) = \begin{cases} 140t^{-1/2} & \text{if } 0 < t < 1 \\ 0.1(t-2)^3(16-(t-2)^4)^{-3/4} & \text{if } 1 < t < 3 \\ -140(4-t)^{-1/2} & \text{if } 3 < t < 4 \end{cases}$$

So, $a(2.5) \approx 0.1$ Km./hour^2.

(c) Using the formula in part (b), we get $a(3.833) \approx -343$ Km./hour^2. The answer is negative because the train is slowing down.

(d) At $t = 3:59:30$, the time is $3 + 59/60 + 30/(60 \times 60)$ hours ≈ 3.99 hours. $v(3.99) \approx 28$ Km./hour and $a(3.99) \approx -1400$ Km./hour^2.

(e) As $t \to 1$ from the left, $a(t) \to 140$, but as $t \to 1$ from the right $a(t) \to -0.013$ Km./hour^2.

6. For any x between 0 and 1, the base of the rectangle is $2x$ and the height is y. From Exercise 11 Figure 2, $y^2 + (1-x)^2 = (2-2x)^2$ or $y = \sqrt{3}(1-x)$.

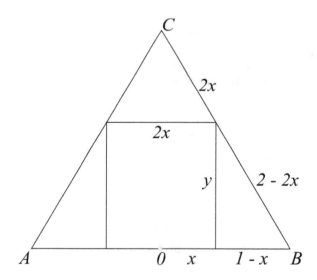

Exercise 11 Figure 2.

The area is $A = 2xy$, or written as a function of x; it is $A(x) = 2\sqrt{3}x(1-x)$. The derivative $A'(x) = 2\sqrt{3}(1-2x)$ is zero at $x = \frac{1}{2}$.

Furthermore, $A''(\frac{1}{2}) < 0$, therefore the area is maximum at $x = \frac{1}{2}$, and it is $A(\frac{1}{2}) = \sqrt{3}/2$.

7. Consider the two functions $g(x)$ and $h(x)$ given below

   (a) $g(x) = -x^4$. The graph of $y = g(x)$ has a maximum at $x = 0$, and both $g'(0)$ and $g''(0)$ are zero.

   (b) $h(x) = x^4$. The graph of $y = h(x)$ has a minimum at $x = 0$, and both $h'(0)$ and $h''(0)$ are zero.

What do you think of that?

12. Higher Derivatives

Answers to Exercises 12

1. If $f(x) = x^4 + x^3 + x^2 + x + 1$,

    (a) $f'(x) = 4x^3 + 3x^2 + 2x + 1$, and $f''(x) = 12x^2 + 6x + 2$.

    (b) The first derivative of $f''(x)$ is $\frac{d}{dx}(12x^2 + 6x + 2) = 24x + 6$; the second derivative of $f''(x)$ is $\frac{d}{dx}(24x + 6) = 24$.

2. We want the third derivative of the product $g(x)h(x)$. In the text, what we have is the second derivative

$$\frac{d^2(g(x)h(x))}{dx^2} = g(x)\frac{dh^2(x)}{dx^2} + 2\frac{dg(x)}{dx}\frac{dh(x)}{dx} + \frac{d^2g(x)}{dx^2}h(x).$$

If we differentiate this

$$\frac{d^3(g(x)h(x))}{dx^3} = \frac{d}{dx}\left(g(x)\frac{d^2h(x)}{dx^2} + 2\frac{dg(x)}{dx}\frac{dh(x)}{dx} + \frac{d^2g(x)}{dx^2}h(x)\right),$$

we get

$$g(x)\frac{d^3h(x)}{dx^3} + 3\frac{dg(x)}{dx}\frac{d^2h(x)}{dx} + 3\frac{d^2g(x)}{dx}\frac{dh(x)}{dx} + \frac{d^3g(x)}{dx}h(x).$$

3. The nth derivative of the product $g(x)h(x)$ is

$$\frac{d^n(g(x)h(x))}{dx^n} = \sum_{k=0}^{k=n} \frac{n!}{(n-k)!k!}\frac{d^kg(x)}{dx^k}\frac{d^{n-k}h(x)}{dx^{n-k}},$$

where $\dfrac{d^0f(x)}{dx^0} = f(x)$.

This may be proved by mathematical induction.

4. Find the derivatives.

    (a) If $y(x) = (x+1)\sin(2x) + x^3 + x^2 + x$, then

$$
\begin{aligned}
\frac{dy(x)}{dx} &= \sin(2x) + 2(x+1)\cos(2x) + 3x^2 + 2x + 1 \\
\frac{d^2y(x)}{dx^2} &= 2\cos(2x) + 2\cos(2x) - 4(x+1)\sin(2x) + 6x + 2, \\
&= 4\cos(2x) - 4(x+1)\sin(2x) + 6x + 2, \\
\frac{d^3y(x)}{dx} &= -8\sin(2x) - 4\sin(2x) - 8(x+1)\cos(2x) + 6, \\
&= -12\sin(2x) - 8(x+1)\cos(2x) + 6.
\end{aligned}
$$

    (b) If $y(x) = \sin^{-1}(x)$, then

$$
\begin{aligned}
y'(x) &= \frac{1}{\sqrt{1-x^2}} = (1-x^2)^{-1/2}, \text{ and} \\
y''(x) &= -\frac{1}{2}(1-x^2)^{-3/2}(-2x) = x(1-x^2)^{-3/2}.
\end{aligned}
$$

5. The function $y(x)$ is given as

$$
y(x) = \begin{cases} x & \text{if } 0 \le x \le 1 \\ -\frac{1}{2}x^2 + 2x - \frac{1}{2} & \text{if } 1 \le x \le 2 \end{cases}.
$$

So, the derivative is

$$
y'(x) = \begin{cases} 1 & \text{if } 0 \le x \le 1 \\ -x + 2 & \text{if } 1 \le x \le 2 \end{cases},
$$

and the graph of $y'(x)$ is as shown in Exercise 12 Figure 1.

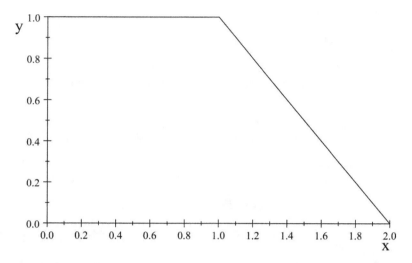

Exercise 12 Figure 1 The graph of $y'(x)$ in Problem 5.

(a) We can see that $y'(1) = 1$ as verified by letting $x \to 1$ from the left. Likewise $y'(1) = 1$, if we use $y'(x) = -x + 2$ and let $x \to 1$ from the right.

(b) If we attempt to find $y''(x)$ at any $x \neq 1$, we get

$$y''(x) = \begin{cases} 0 & \text{if } 0 \leq x < 1 \\ -1 & \text{if } 1 < x \leq 2 \end{cases}.$$

Now as $x \to 1$ from the left, $y''(x) \to 0$, but from the right $y''(x) \to -1$. Therefore $y''(1)$ does not exist.

13. Continuous Functions

Answers to Exercises 13

1. The function is

$$y_2(x) = \begin{cases} -\sqrt{x} & \text{if} \quad 0 < x \le 2 \\ \sqrt{x} & \text{if} \quad 2 < x \le 5 \end{cases}.$$

(a) The graph is

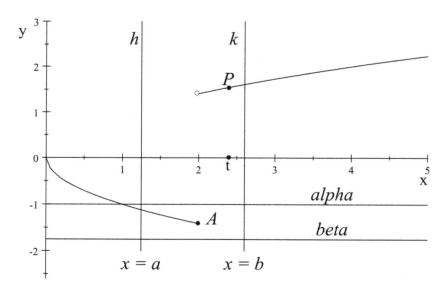

Exercise 13 Figure 1 Graph of $y_2(x)$.

(b) The point $A = (2, \sqrt{2})$. The graph is not continuous at A because there exist two horizontal lines α and β with A between them such that for any two vertical lines h and k with A between them there will always be some point of the graph between h and k that is not between α and β.

Here is the proof of that.   Let α be the horizontal line whose equation is $y = -1$ and β be the line whose equation is $y = -1.75$. See Exercise 13 Figure 1.   Since the ordinate of A is $-\sqrt{2}$ which is between -1 and -1.75, then A is between α and β.   Let h and k be any two vertical lines, whose equations are $x = a$ and $x = b$, respectively and such that $0 < a < 2$ and $2 < b \leq 5$.   The point A is between h and k.   Now, since $b > 2$, there is some number t between $x = 2$ and $x = b$ such that $P = (t, \sqrt{t})$ is on the graph, between h and k but, not between α and β because $\sqrt{t} > 0$.

2. The domain D consists of the single number $x = 0$.   The equation is $s(x) = x + 1$, so the entire graph consists of the single point $A = (0, 1)$.

(a) Here is the graph.

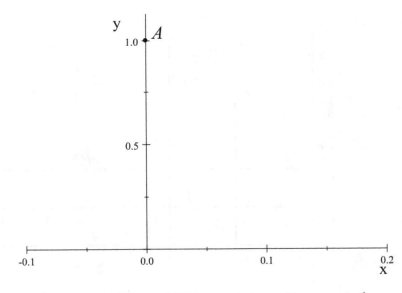

Exercise 13 Figure 2 This graph is continuous at A.

(b) This graph is continuous at its only point.   Otherwise, there would be two horizontal lines α and β with A between them such that every pair of vertical lines h and k with A between them would have a point of the graph between h and k and not between α

and β. Such a point would be distinct from A. But there are no points of the graph distinct from A.

3. The function $w(x)$ is

$$w(x) = \begin{cases} \sin(x) & \text{if } 0 < x < \pi/2 \\ 0 & \text{if } \pi/2 \leq x \end{cases}.$$

The function is not continuous at the point whose abscissa is $\pi/2$. Let $A = (\pi/2, 0)$ be that point. If α is the horizontal line with equation $y = 1/4$ and β is the horizontal line with equation $y = -1/4$, the point A is between α and β. See Exercise 13 Figure 3.

Let $x = h$ and $x = k$ be any two vertical lines with A between them; that is, $h < \pi/2 < k$. Since $h < \pi/2$, there is some number $t < \pi/2$ and greater than the maximum of h and $\pi/4$, therefore $\sin(t)$ is greater than $\sin(\pi/4)$, that is $\sin(t) > 0.707$. If we let $P = (t, \sin(t))$, then P is a point of the graph between $x = h$ and $x = k$, that is not between α and β. Hence the graph is not continuous at A.

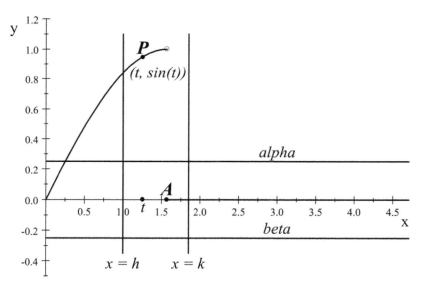

Exercise 13 Figure 3 The graph is not continuous at A.

4. The function $g(x)$ is

$$g(x) = \begin{cases} \frac{x^2-1}{x-1} & \text{if } x \neq 1 \\ 2 & \text{if } x = 1 \end{cases}.$$

(a) The graph of $g(x)$ on $-2 \leq x \leq 3$ is shown in Exercise 13 Figure 4.

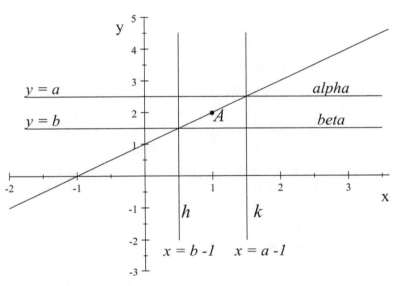

Exercise 13 Figure 4 Graph of $y = g(x)$.

(b) Yes, the graph is continuous at $A = (1, 2)$. If we look at the equation for $g(x)$ for any $x \neq 1$, we see that $\frac{x^2-1}{x-1} = x + 1$, and that for $x = 1$, $g(1) = 2$. Let α and β be any two horizontal lines with equations $y = a$ and $y = b$, $a > b$, respectively and with A between them. Then $a > 1 > b$ and the vertical lines h and k with equations $x = b - 1$ and $x = a - 1$ will have A between them and every point of the graph between h and k will be between α and β. Therefore, the graph of $y = g(x)$ is continuous at A.

5. The graph of $y = x^2$ is continuous at the point $A = (1, 1)$ because if α and β are any two horizontal lines whose equations are $y = a$, $y = b$, $a > 1 > b$, with A between them, then there exist two vertical lines, h and k whose equations are $x = \sqrt{b}$, and $x = \sqrt{a}$, with A between them such that every point of the graph between h and k is also between α and β. This is so because for every x such that $\sqrt{b} < x < \sqrt{a}$, we have $b < x^2 < a$.

6. The given function is

$$m(x) = \begin{cases} 0 & \text{if} \quad x \text{ is a rational number} \\ 1 & \text{if} \quad x \text{ is an irrational number} \end{cases}.$$

(a) $m(5/3) = 0$, because $5/3$ is rational.

(b) $m(\pi) = 1$, because π is irrational.

(c) m is not continuous at $x = 1$.

Here is a proof. Let A be the point of m whose abscissa is 1. Then $A = (1, 0)$. Now there exist horizontal lines $y = 1/2$ and $y = -1/2$ with A between them, such that if $x = h$ and $x = k$ are any two vertical lines with A between them, there is always some point B of m on the horizontal line $y = 1$ between $x = h$ and $x = k$.

This is true because between the number h and the number k, there is always an irrational number t and $m(t) = 1$, which is not between $1/2$ and $-1/2$.

14. Functions Continuous on Their Domains.

Answers to Exercises 14

1. Let G be the graph of $y = x^2$ on D, the rationals in the interval $[0, 1]$. Let the point $A = (m/n, (m/n)^2)$ be any point of G. Let α and β be horizontal lines with equations $y = a^2$ and $y = b^2$, respectively, with A between them. So, $0 \le b^2 < (m/n)^2 < a^2 \le 1$. Now we can find two rational numbers p/q and r/s such that $b^2 < (r/s)^2 < (m/n)^2 < (p/q)^2 < a^2$. For example, between the positive rational number m/n and any number $a > m/n$, we can find a rational number p/q. Thus,

$$0 < \frac{m}{n} < \frac{p}{q} < a.$$

Therefore

$$\frac{m^2}{n^2} < \frac{mp}{nq}.$$

Similarly

$$\frac{mp}{nq} < \frac{p^2}{q^2}.$$

Therefore

$$\frac{m^2}{n^2} < \frac{p^2}{q^2}.$$

In the same way we can prove that

$$\frac{p^2}{q^2} < a^2.$$

349

These, and other similar inequalities tell us that for any two horizontal lines $y = (p/q)^2$ and $y = (r/s)^2$ with $A = (m/n, (m/n)^2)$ between them, there exist two vertical lines $x = p/q$ and $x = r/s$ with A between them such that every point of G between the vertical lines is also between the horizontal lines. Thus, the graph G is continuous at A.

2. Here, the function $g(x)$ is

$$g(x) = \begin{cases} x^2 & \text{if} \quad x \text{ is rational} \\ 0 & \text{if} \quad x \text{ is irrational} \end{cases}.$$

This time, we want to prove that the function is not continuous for any x except $x = 0$. We will do this by drawing graphs to illustrate the discontinuity of G, the graph of the equation $y = g(x)$. The rational points (the ones with rational abscissas) of G are on the parabola $y = x^2$. The irrational points of G (the ones with irrational abscissas) are on the x-axis, that is $y = 0$.

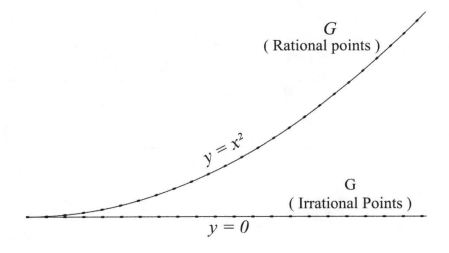

Exercise 14, Figure 1 The graph G in Problem 2.

Now Suppose $A = (a, 0)$ is any irrational point of G. See Exercise 14, Figure 2. In this figure there exist two horizontal α and β with A between them such that if h and k are any two vertical lines with A

between them, there will always be some rational points of G between h and k that are not between α and β. So G is discontinuous at A.

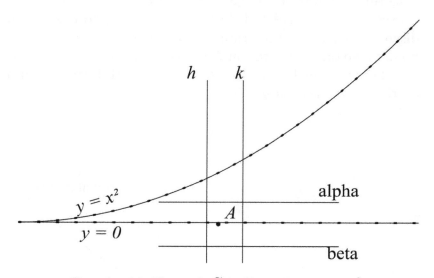

Exercise 14, Figure 2 G is discontinuous at A.

Similarly, the graph can be shown to be discontinuous at any point P which has a nonzero rational abscissa. See Exercise 14, Figure 3. The graph is continuous, however, at the point $(0,0)$, where $x^2 = 0$.

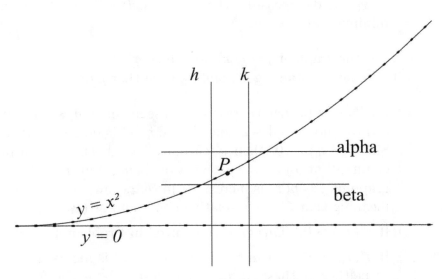

Exercise 14, Figure 3 G is discontinuous at P.

3. We are given that G is the graph of a continuous function $y = g(x)$ on a segment (a, b) containing the point x_0. If $g(x_0) > 0$, and $A = (x_0, g(x_0))$, let α and β be two horizontal lines with A between them such that α is above A and the equation of β is $y = \frac{1}{2}g(x_0)$. Now since G is continuous at A, there exist vertical lines h, equation $x = b$, and k, equation $x = d$, $c < d$, with A between them such that every point of G between h and k is between α and β, therefore above the line $y = \frac{1}{2}g(x_0)$, which is positive. Hence $g(x) > 0$ in the segment (b, d).

4. Let M be the graph of $y = m(x)$.

    (a) We have that $m'(x_0) = 0$ and $m''(x_0) > 0$ at some point of (a, b), and we want to show that $m''(x) > 0$ on a subsegment of (a, b). Since $m''(x)$ is continuous on a segment (a, b) containing x_0 and $m''(x_0) > 0$, then by Problem 3, there is some segment (b, d) such that $m''(x) > 0$ for all x in (b, d).

    (b) We want to show that the graph of $y = m(x)$ has a relative minimum at $(x_0, m(x_0)$. The fact that $m''(x)$ is positive throughout (b, d) means that $m'(x)$ is increasing throughout it. But $m'(x_0) = 0$ so $m'(x) < 0$ for all $x \in (b, x_0)$ and $m'(x) > 0$, when $x \in (x_0, d)$. This means that M, itself is decreasing while x is in

(b, x_0), and increasing when $x \in (x_0, d)$. Hence M has a relative minimum at $(x_0, m(x_0))$.

5. Let G be the graph of $y = g(x)$, for which $g''(x)$ exists and is continuous on its domain containing a point x_0 such that $g'(x_0) = 0$.

   (a) If $g''(x_0) < 0$, then there exists an segment (b, d) such that $g''(x) < 0$ throughout that segment, hence $g'(x)$ is decreasing throughout. Since $g'(x_0) = 0$, then $g'(x)$ must have been positive before becoming 0 at x_0 and negative after being 0 at x_0. Therefore the graph of $y = g(x)$ is increasing in (b, x_0) and decreasing in (x_0, d) meaning that G has a relative maximum at $(x_0, g(x_0))$.

   (b) If $g''(x_0) > 0$. This case was solved in Problem 4.

   (c) If $g''(x_0) = 0$, we can't tell whether $g'(x)$ is increasing, decreasing or neither of these at x_0. The test fails as the following three counter-examples show.

      i. If $g(x) = x^4$, then $g'(0) = 0$, $g''(0) = 0$, and the graph of $y = g(x)$ is minimum at 0.
      ii. If $g(x) = -x^4$, then $g'(0) = 0$, $g''(0) = 0$, and the graph of $y = g(x)$ is maximum at 0.
      iii. If $g(x) = x^3$, then $g'(0) = 0$, $g''(0) = 0$, and the graph of $y = g(x)$ has an inflection point at 0.

15. Some Maximum - Minimum Problems

Answers to Exercises 15

1. Given the line L with equation $Ax + By + C = 0$, $A^2 + B^2 \neq 0$, and any point $P = (p, q)$ in the plane, we will consider two cases: $P \in L$ and $P \notin L$.

Case 1. $P \in L$, then $Ap + Bq + C = 0$, so

$$s = \left| \frac{Ap + Bq + C}{\sqrt{A^2 + B^2}} \right| = 0.$$

Thus s, the distance to the line from a point on the line, is zero.

Case 2. $P \notin L$. Here we will consider three sub-cases: First $A = 0$, and $B \neq 0$, second $A \neq 0$ and $B = 0$, third neither A nor B is zero. First, if $A = 0$, the line is horizontal and has equation $y = -C/B$, the distance of any point (p, q) from this line is $|q + C/B|$, therefore

$$s = \left| \frac{Bq + C}{B} \right|, \text{ or}$$

$$s = \left| \frac{0p + Bq + C}{\sqrt{0^2 + B^2}} \right|.$$

Second, if $B = 0$, the line is vertical and has equation $x = -C/A$, the distance of any point (p, q) from this line is $|p + C/A|$, therefore

$$s = \left| \frac{Ap + C}{A} \right|, \text{ or}$$

$$s = \left| \frac{Ap + 0q + C}{\sqrt{A^2 + 0^2}} \right|.$$

355

Third, if neither A nor B is zero, then the distance from P to any point (x, y) on L is

$$s = \sqrt{(x - p)^2 + (y - q)^2},$$

but y is actually a function of x, that is $y = y(x) = -(A/B)x - (C/B)$. This makes s a function of x,

$$s(x) = \sqrt{(x - p)^2 + (y(x) - q)^2}.$$

To find the shortest distance from P to L, we find the derivative and set it equal to zero getting

$$(x - p) + (y - q)(-A/B) = 0, \text{ or}$$
$$(y - q) = (B/A)(x - p). \tag{I}$$

Let $x = x_0$ be the solution to this equation, then

$$x_0 = \frac{B^2 p - AC - ABq}{A^2 + B^2}, \tag{II}$$

and

$$y_0 = (B/A)(x_0 - p) + q.$$

Then the distance becomes

$$s(x_0) = \sqrt{(x_0 - p)^2 + \frac{B^2}{A^2}(x_0 - p)^2},$$
$$= \left| \frac{x_0 - p}{A} \right| \sqrt{A^2 + B^2}. \tag{III}$$

Now plug the value of x_0 from Equation (II) into Equation (III) and you get

$$s(x_0) = \left| \frac{Ap + Bq + C}{\sqrt{A^2 + B^2}} \right|.$$

2. Let G be the graph of the equation

$$y = 2\sqrt{x - 1} + 1, \text{ for } x \geq 1.$$

If A is the point $(4, 1)$, find the distance from A to G. See the graph in Exercise 15 Figure 1, which shows both the point A for this problem and the point B for problem 3.

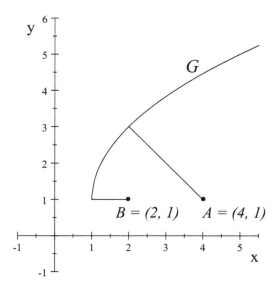

Exercise 15 Figure 1 Distance from A to G and Distance from B to G.

The formula for the distance from A to any point (x, y) on the graph G is

$$
\begin{aligned}
s(x) &= \sqrt{(x-4)^2 + (y-1)^2}, \\
s(x) &= \sqrt{(x-4)^2 + 4(x-1)}, \text{ and} \\
s'(x) &= \frac{2(x-4)+4}{2\sqrt{(x-4)^2 + 4(x-1)}}.
\end{aligned}
$$

Now solving $s'(x_0) = 0$, we get $x_0 = 2$. The distance from A to G is $\sqrt{8}$.

3. Again, we are using the graph of $y = 2\sqrt{x-1} + 1$ with $x \geq 1$. The formula for the distance from point $B = (2, 1)$ to G is

$$
\begin{aligned}
s(x) &= \sqrt{(x-2)^2 + 4(x-1)}, \text{ and} \\
s'(x) &= \frac{x}{\sqrt{x^2}} = \frac{x}{|x|}.
\end{aligned}
$$

But this value cannot be zero. Since $x \geq 1$, the minimum value of $s(x)$ occurs at the end point, $x = 1$. Therefore, the distance from B to G is 1.

4. The intercept form of a line equation is

$$\frac{x}{a} + \frac{y}{b} = 1.$$

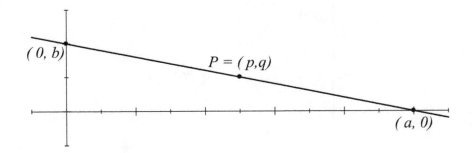

Exercise 15 Figure 2. Find the triangle of minimum area.

If (p, q) is on the line we get $\frac{p}{a} + \frac{q}{b} = 1$ implying

$$b = \frac{aq}{a - p}.$$

The area $A = \frac{1}{2}a \times b$, so area as a function of a is

$$A(a) = \frac{1}{2}\frac{a^2 q}{a - p}.$$

The derivative of $A(a)$ is

$$A'(a) = \frac{1}{2}\frac{aq(a - 2p)}{(a - p)^2}.$$

So, $A'(a) = 0$ at $a = 2p$, and consequently $b = 2q$. If we check $A''(a)$, we will see that $A''(2p) > 0$. Therefore the line through (p, q) that creates the minimum possible area is

$$\frac{x}{2p} + \frac{y}{2q} = 1.$$

And that minimum area is $A(2p) = 2pq$.

5. What is the shortest ladder that can reach across a fence q feet high and p feet from the house? See Exercise 15 Figure 3.

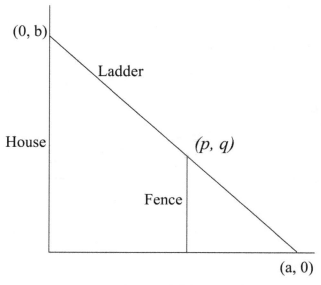

Exercise 15 Figure 3.

Again, we use the intercept form of the line equation $\frac{x}{a} + \frac{y}{b} = 1$. If (p, q) is a point on this line then, as before, we get

$$b = \frac{aq}{a - p},$$

but, this time we want to minimize L, the length of the ladder.

$$L = \sqrt{a^2 + b^2}.$$

Writing this as a function of a,

$$L(a) = \sqrt{a^2 + \left(\frac{aq}{a - p} \right)^2}.$$

First find $L'(a)$, then what value of a makes $L'(a) = 0$. We end up with

$$(a - p)^3 - pq^2 = 0.$$

If we solve this equation for a, we get

$$a = p + p^{1/3}q^{2/3},$$

so

$$a = p^{1/3}(p^{2/3} + q^{2/3}), \text{ and}$$
$$b = q^{1/3}(p^{2/3} + q^{2/3}).$$

Computing $L''(a)$ we can see that it is positive; therefore, the shortest ladder over the fence is

$$L = \sqrt{(p^{1/3}(p^{2/3} + q^{2/3}))^2 + (q^{1/3}(p^{2/3} + q^{2/3}))^2}, \text{ or}$$
$$L = (p^{2/3} + q^{2/3})^{3/2}.$$

6. Two corridors of widths p and q meet at a right angle.

(a) The equation of the line is $bx + ay - ab = 0$, and the length of the ladder is $L = \sqrt{a^2 + b^2}$. If the point $P = (p, q)$, is not on the line then by Problem 1 in this Exercise, the distance from P to the line is

$$s(a) = \left| \frac{bp + aq - ab}{\sqrt{a^2 + b^2}} \right| > 0.$$

But any ladder that never touches the corner is not the longest one that can pass through. So we need to find the shortest ladder that touches the corner and both the x-axis and the y-axis. The shortest one of these satisfies the condition that $s(a_0) = 0$, when $L'(a_0) = 0$. Thus, we get from the two equations

$$b = \frac{aq}{a - p}, \text{ and}$$
$$L(a) = \sqrt{a^2 + b^2},$$

therefore,

$$a = p^{1/3}(p^{2/3} + q^{2/3}), \text{ and}$$
$$b = q^{1/3}(p^{2/3} + q^{2/3})$$

as we found in Problem 5. The length of the shortest ladder satisfying these conditions is

$$L = (p^{2/3} + q^{2/3})^{3/2},$$

which is the longest ladder that can go around the corner.

(b) Two corridors whose widths are p feet and q feet $(p > 0, q > 0)$ are right angle to each other. We want to find the length of the longest ladder that can be carried horizontally around the corner from one corridor into the next. Let us consider the set of all ladders that are too long to carry around the corner, and notice that there are some ladders short enough to be carried around. This means that the too long ladders have a lower bound; therefore there must be a *shortest ladder* that is too long. When we solved the "ladder over the fence" problem we found the shortest ladder touching the fence, the ground and the house. A ladder of this length will also be the shortest ladder that can't get around the corner.

7. At any time t, let $x(t)$ be the distance traveled by the East-going car and $y(t)$ be the distance traveled by the North-going car. Therefore, their distance apart at any time t is

$$s(t) = \sqrt{x(t)^2 + y(t)^2}.$$

The data we are given can be written as

$$
\begin{aligned}
x(0) &= 4, \\
x'(0) &= 60, \\
y(0) &= 5, \\
y'(0) &= 40.
\end{aligned}
$$

The rate of change in the distance at any time t is $s'(t)$,

$$s'(t) = \frac{2x(t)x'(t) + 2y(t)y'(t)}{2\sqrt{x(t)^2 + y(t)^2}},$$

so at time $t = 0$, we get the rate at which their distance apart is increasing $s'(0)$, namely

$$
\begin{aligned}
s'(0) &= \frac{x(0)x'(0) + y(0)y'(0)}{\sqrt{x(0)^2 + y(0)^2}} \\
&= \frac{4 \times 60 + 5 \times 40}{\sqrt{4^2 + 5^2}}, \\
&= \frac{440}{\sqrt{41}} \approx 69 \text{ m.p.h.}
\end{aligned}
$$

16. Max-Min Problems in Geometry

Answers to Exercises 16

1. Given the volume of the cylinder is 4 cubic inches, and we want the minimum surface area. The formulas for the volume, V, and surface area, S, are as follows:

$$V = \pi r^2 h, \text{ and}$$
$$S = 2\pi rh + 2\pi r^2,$$

where r is the radius and h is the height. From the given volume, we have $4 = \pi r^2 h$, so $h = 4/(\pi r^2)$, making the surface area the following function of r

$$S(r) = \frac{8}{r} + 2\pi r^2.$$

The derivative is

$$S'(r) = -\frac{8}{r^2} + 4\pi r.$$

Setting $S'(r_0) = 0$, we get

$$r_0 = \left(\frac{2}{\pi}\right)^{1/3}.$$

Since

$$h = \frac{4}{\pi r^2},$$

we have

$$h_0 = \frac{4}{\pi r_0^2} = \left(\frac{4}{\pi}\right)\left(\frac{\pi}{2}\right)^{2/3},$$

363

or

$$h_0 = 2 \left(\frac{2}{\pi}\right)^{1/3}.$$

Thus, the height of the cylinder is equal to 2 times r_0, the diameter of the base, meaning that the cylinder has a minimum surface area when it has a square silhouette. The area is

$$S = 8 \left(\frac{\pi}{2}\right)^{1/3} + 2\pi \left(\frac{2}{\pi}\right)^{2/3} \approx 13.95 \text{ in}^2.$$

2. To solve the problem of finding the maximum volume for a cylindrical can whose total surface area is a fixed number, let r be the radius, and h be the height. Let $S = A$ be the fixed given area, then

$$\begin{aligned} A &= 2\pi rh + 2\pi r^2, \\ V &= \pi r^2 h. \end{aligned}$$

Solve the first equation for h,

$$h = \frac{A - 2\pi r^2}{2\pi r}.$$

Then write the volume equation as a function of r as follows

$$V(r) = \pi r^2 h = \frac{r}{2}(A - 2\pi r^2).$$

The derivative, $V'(r) = \frac{A}{2} - 3\pi r^2$. So $V'(r_0) = 0$, when r_0 is

$$r_0 = \sqrt{\frac{A}{6\pi}}.$$

Now substituting into the equation for h, we get

$$\begin{aligned} h_0 &= \sqrt{\frac{2A}{3\pi}}, \text{ or} \\ h_0 &= 2r_0. \end{aligned}$$

So, again the height is the same as the diameter, making the can have a square silhouette. The maximum volume is

$$V(x_0) = \pi r_0^2 h_0 = 2\pi r_0^3 = \frac{A}{3}\sqrt{\frac{A}{6\pi}}.$$

3. We want to find the foci, F_1 and F_2 of the ellipse whose equation is

$$\frac{x^2}{16} + \frac{y^2}{9} = 1.$$

First we need to find the number k such that the sum of the distances from any point on the ellipse to the two foci is k. Let Q be the point $(0,3)$. Q satisfies the equation and is on the ellipse. Also Q is on the minor axis so its distance from F_1 is the same as its distance from F_2. Let the coordinates of the focus F_1 be $(a,0)$ and $F_2 = (-a,0)$. The distance $\overline{QF_1} = \overline{QF_2} = \sqrt{a^2 + 3^2}$. The sum of these two distances is $2\sqrt{a^2 + 9}$ The point $R = (4,0)$ is on the major axis and its distance from the two foci are $\overline{RF_1} = 4 - a$ and $\overline{RF_2} = 4 + a$. The sum of these two distances is 8. So, $2\sqrt{a^2 + 9} = 8$, solving for a, we get $a = \sqrt{7}$. The foci are $(\sqrt{7}, 0)$ and $(-\sqrt{7}, 0)$.

4. The area of the square with base x is x^2 and the area of the circle with diameter $1 - x$ is $\pi(\frac{1-x}{2})^2$. The variable x is restricted to $0 \le x \le 1$. The area function $A(x)$ is

$$A(x) = x^2 + \pi\left(\frac{1-x}{2}\right)^2.$$

The first derivative $A'(x) = (2 + \frac{\pi}{2})x - \frac{\pi}{2}$, which is zero when $x_0 = \frac{\pi}{\pi + 4}$, but at this point $A''(x_0) > 0$; therefore, A is minimum at x_0. The maximum area occurs at the end point $x = 1$.

5. The area, $A(x)$, of the given quadrilateral is made up of the rectangle with base 1 and height x plus the two right triangles each with height x and hypotenuse 1 making the base $y = \sqrt{1 - x^2}$. Thus,

$$\begin{aligned} A(x) &= x + x\sqrt{1 - x^2}, \\ A'(x) &= 1 + \sqrt{1 - x^2} - \frac{x^2}{\sqrt{1 - x^2}}. \end{aligned}$$

Finding x_0 so that $A'(x_0) = 0$, we get $x_0 = \sqrt{3}/2$. Now compute $A''(x)$, getting

$$\begin{aligned} A''(x) &= -\frac{x^3}{(1 - x^2)^{3/2}} - \frac{3x}{(1 - x^2)^{1/2}}, \text{ so} \\ A''(x_0) &= -6\sqrt{3}, \end{aligned}$$

which is negative; so the area is maximum at x_0, and the maximum area is

$$A(\frac{\sqrt{3}}{2}) = \frac{3\sqrt{3}}{4} \approx 1.299.$$

17. The Derivative of $\log_b(x)$

Answers to Exercises 17

1. If $y(x) = 10^x$, then $\log(y(x)) = x$. Therefore, $\frac{d\log(y(x))}{dx} = 1$.

2. If $y(x) = \log_5(x^2 + x + 1)$, then

$$y'(x) = \frac{2x + 1}{x^2 + x + 1} \log_5(e).$$

3. If $g(x)$ is differentiable and $\log_7(g(x))$ exists then

$$\frac{d\log_7(g(x))}{dx} = \frac{g'(x)}{g(x)} \log_7(e).$$

4. If $y(x) = 10^x$, then we can find $y'(x)$ by using Problem 1,

$$\frac{d\log(y(x))}{dx} = 1,$$

and by the chain rule

$$
\begin{aligned}
\frac{d\log(y(x))}{dx} &= \frac{d\log(y(x))}{dy(x)} \frac{dy(x)}{dx} \log(e) \\
&= \frac{y'(x)}{y(x)} \log(e).
\end{aligned}
$$

So,

$$\frac{y'(x)}{y(x)} \log(e) = 1,$$

or

$$y'(x) = \frac{y(x)}{\log(e)} = \frac{10^x}{\log(e)}, \text{ or}$$
$$y'(x) = 10^x \ln(10).$$

5. To find
$$\frac{d(e^x)}{dx},$$
we can start by letting $y(x) = e^x$, and then find $y'(x)$, thus,

$$y(x) = e^x$$
$$\ln(y(x)) = x,$$
$$\frac{d}{dx}\ln(y(x)) = \frac{y'(x)}{y(x)} = 1.$$

So,
$$y'(x) = y(x) = e^x.$$

6. If
$$y(x) = \ln(e^{\sin(x^2)}),$$
first, let's write
$$\ln(e^{\sin(x^2)}) = \sin(x^2)\ln(e) = \sin(x^2).$$

Therefore,
$$y(x) = \sin(x^2),$$
use the chain rule to get
$$y'(x) = \cos(x^2)2x.$$

7. If

$$y(x) = \sin(\ln(e^{x^2})), \text{ then}$$
$$y(x) = \sin(x^2), \text{ and}$$
$$y'(x) = \cos(x^2)2x.$$

8. If
$$y(x) = \log(\sin^{-1}(\log(x))),$$

By the chain rule

$$y'(x) = \frac{d\log(\sin^{-1}(\log(x)))}{d\sin^{-1}(\log(x))} \frac{d\sin^{-1}(\log(x))}{d\log(x)} \frac{d\log(x)}{dx}$$

$$y'(x) = \frac{1}{\sin^{-1}(\log(x))}\log(e)\frac{1}{\sqrt{1-\log^2(x)}}\frac{1}{x}\log(e).$$

But this answer may be written in other equivalent forms. For example we note that for $x > 0$, $\log(x) = \frac{\ln(x)}{\ln(10)}$, and $\log(e) = \frac{1}{\ln(10)}$. In other words, we can say

$$y'(x) = \frac{1}{x\sin^{-1}[\ln(x)/\ln(10)]\ln^2(10)\sqrt{1-\ln^2(x)/\ln^2(10)}}.$$

9. If $y(x) = \log\sqrt{1-x^2}$, then $y = \frac{1}{2}\log(1-x^2)$, so

$$y'(x) = \frac{1}{2}\frac{-2x}{1-x^2}\log(e),$$

$$= \frac{-x}{1-x^2}\log(e).$$

10. The derivatives

   (a) If $y = \ln(\cos(x))$, then $y'(x) = \frac{-\sin(x)}{\cos(x)} = -\tan(x)$

   (b) If $y = \ln(\sec(x))$, then $y(x) = \ln(\frac{1}{\cos(x)}) = -\ln(\cos(x))$, so $y'(x) = \tan(x)$.

11. To show how the indeterminate form 1^∞ could stand for π, find

$$\lim_{n\to\infty} (1 + \frac{1}{n})^{n\ln(\pi)}.$$

If we just let $n\ln(\pi) \to \infty$ and $\frac{1}{n} \to 0$, we would have the indeterminate form 1^∞. But if we first rewrite the problem as

$$\lim_{n\to\infty} \left((1 + \frac{1}{n})^n\right)^{\ln(\pi)},$$

we get

$$e^{\ln(\pi)} = \pi.$$

12. If $x > 0$ and $y(x) = x^a$, then

$$\begin{aligned}
\ln(y(x)) &= a\ln(x), \\
\frac{y'(x)}{y(x)} &= \frac{a}{x}, \text{ so} \\
y'(x) &= \frac{a}{x}y(x) = ax^{a-1}.
\end{aligned}$$

13. If $b > 0$ and $b \neq 1$, and $y(x) = b^x$, then $\ln(y(x)) = x\ln(b)$, so

$$\begin{aligned}
\frac{y'(x)}{y(x)} &= \ln(b) \\
y'(x) &= y(x)\ln(b), \\
y'(x) &= b^x\ln(b).
\end{aligned}$$

14. If

$$y(x) = 1 + x + \frac{x^2}{2!} + \frac{x^3}{3!} + \frac{x^4}{4!} + \ldots + \frac{x^n}{n!} + \ldots,$$

then

$$y'(x) = 1 + x + \frac{x^2}{2!} + \frac{x^3}{3!} + \frac{x^4}{4!} + \ldots + \frac{x^n}{n!} + \ldots.$$

18. The Function $\sin\left(\frac{1}{x}\right)$

Answers to Exercises 18

1. The graph of

$$h(x) = \begin{cases} x\sin(\frac{1}{x}) & \text{if } x \neq 0 \\ 0 & \text{if } x = 0 \end{cases},$$

is shown in Exercise 18, Figure 1.

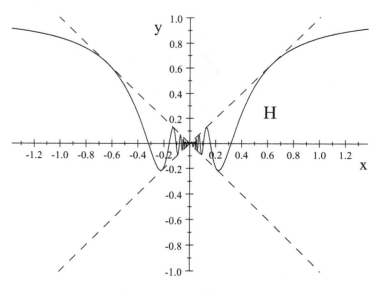

Exercise 18, Figure 1.

(a) We know that $P = (0,0)$ is a point of the graph   This is the first condition necessary for H to be continuous at P. It is not, however, sufficient. Before we continue with the proof that the

371

graph is continuous at $(0,0)$, let us note that, for any real number $x \neq 0$,

$$-x \leq x \sin(1/x) \leq x.$$

Now, let α and β be any two horizontal lines with P between them, and let $y = -c$ and $y = c$, $c > 0$, be two other horizontal lines between α and β also having P between them. These two horizontal lines intersect the lines $y = x$ and $y = -x$ at the four points: (c, c), $(c, -c)$, $(-c, c)$, and $(-c, -c)$. Let h and k be two vertical lines with equations $x = -c$, and $x = c$, respectively. If Q is any point of the graph between the vertical lines h and k, and q is the abscissa of Q, then $-c \leq q \leq c$. Therefore the ordinate of Q satisfies the inequality

$$-c \leq -q \leq q \sin(1/q) \leq q \leq c.$$

Thus, Q is between α and β. This proves H is continuous at the origin.

(b) The graph H does not have a derivative at $x = 0$. We see this by the definition of derivative as follows.

$$
\begin{aligned}
h'(0) &= \lim_{x \to 0} \frac{h(x) - h(0)}{x - 0} \\
&= \lim_{x \to 0} \frac{x \sin(1/x) - 0}{x - 0}, \\
&= \lim_{x \to 0} \sin(1/x), \text{ which oscillates between } -1 \text{ and } +1.
\end{aligned}
$$

This says that the limit does not exist; therefore $h'(x)$ does not exist at $x = 0$. For any other x, the derivative is

$$h'(x) = \sin(\frac{1}{x}) - \frac{1}{x} \cos(\frac{1}{x}).$$

2. If $a > 0$ is any real number and $f(x)$ is defined as

$$f(x) = \begin{cases} \sin(\frac{1}{x}) & \text{if } x \neq 0 \\ a & \text{if } x = 0 \end{cases},$$

the graph F of the function $f(x)$ cannot be continuous at $P = (0, a)$. Here is a proof. Let α and β be two horizontal lines whose equations are $y = c_1$ and $y = c_2$, respectively, and assume $0 < c_2 < a < c_1$, thus $(0, a)$ is between α and β. See Exercise 18, Figure 2.

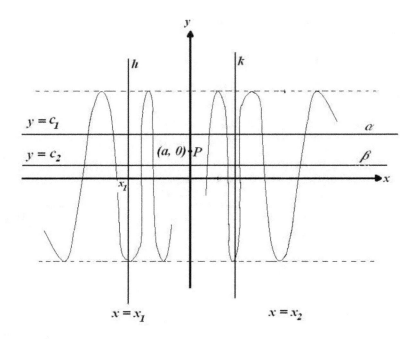

Exercise 18, Figure 2.

Now, for any two vertical lines, h and k with equations $x = x_1$ and $x = x_2$, where $x_1 < 0 < x_2$, with P between them there exist points of F between h and k that are not between α and β because there are integers n such that

$$0 < \frac{2}{(4n-1)\pi} < x_2.$$

Thus a point of F with this abscissa would have for its ordinate, $\sin((4n-1)\pi/2)$ which is -1, and not between α and β. A similar proof can be given for $a < 0$.

3. The graph of

$$h(x) = \begin{cases} x^4 \sin(1/x) & \text{if } x \neq 0 \\ 0 & \text{if } x = 0 \end{cases},$$

is shown in Exercise 18, Figure 3. (This figure is a cartoon; that is it has been manipulated to make the graph more visible.)

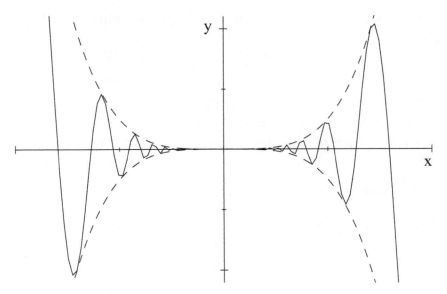

Exercise 18 Figure 3, $h(x) = x^4 \sin(1/x)$.

(a) The graph is continuous at $P = (0,0)$. Any two horizontal lines $y = a$ and $y = -a$ will intersect the graph of $y = x^4$ and $y = -x^4$ in the four points: $(\sqrt[4]{a}, a), (\sqrt[4]{a}, -a), (-\sqrt[4]{a}, a), (-\sqrt[4]{a}, -a)$. Therefore each point of the graph whose abscissa, x, is between $-\sqrt[4]{a}$ and $\sqrt[4]{a}$, will have as its ordinate $x^4 \sin(1/x)$, which will be between a and $-a$. That is, if

$$-\sqrt[4]{a} \;<\; x < \sqrt[4]{a}, \text{ then}$$
$$-a \;<\; x^4 \sin(1/x) < a.$$

Therefore, the graph is continuous at $P = (0,0)$.

(b) The derivative at $x = 0$ is

$$
\begin{aligned}
h'(0) &= \lim_{x \to 0} \frac{h(x) - h(0)}{x - 0} \\
&= \lim_{x \to 0} \frac{x^4 \sin(1/x) - 0}{x - 0}, \\
&= \lim_{x \to 0} x^3 \sin(1/x) = 0.
\end{aligned}
$$

Now, when $x \neq 0$, we have

$$h'(x) = 4x^3 \sin(1/x) - x^2 \cos(1/x).$$

4. Here is a repeat of the derivative found in Equation (18.6).

$$r'(x) = \begin{cases} 2x\sin(\frac{1}{x}) - \cos(\frac{1}{x}) & \text{if } x \neq 0 \\ 0 & \text{if } x = 0 \end{cases}.$$

This function has two terms when $x \neq 0$. The first one, $2x\sin(1/x)$ is continuous at $x = 0$, but as $x \to 0$, $\cos(1/x)$ oscillates between $+1$ and -1 so it does not approach zero. Therefore, $r'(x)$ is not continuous at 0.

5. By definition of the derivative at $x = 0$,

$$\begin{aligned} f'(0) &= \lim_{x \to 0} \frac{x^p \sin(1/x) - 0}{x - 0}, \\ &= \lim_{x \to 0} x^{p-1} \sin(1/x). \end{aligned}$$

As $x \to 0$, $x^{p-1} \to \infty$, because $p - 1 < 0$ and $\sin(1/x)$ oscillates between -1 and $+1$, so the derivative of f does not exist at $x = 0$.

6. Again, as in Problem 5

$$f'(0) = \lim_{x \to 0} x^{p-1} \sin(1/x),$$

but this time $p - 1 > 0$, and $x^{p-1} \to 0$, as $x \to 0$. So, $f'(0)$ does exist and it is zero.

7. If $p > 2$, then,

$$f'(x) = px^{p-1} \sin(1/x) - x^{p-2} \cos(1/x),$$

and when $p > 2$, both $p - 1$ and $p - 2$ are positive so $f'(x) \to 0$ as $x \to 0$. Therefore $f'(x)$ is continuous at 0.

8. When $1 < p < 2$, then $p - 2 < 0$ and the term

$$x^{p-2} \cos(1/x)$$

does not have a limit as $x \to 0$. Therefore, $f'(x)$ is not continuous at 0.

19. Antiderivatives of a Function

Answers to Exercises 19

1. Find an antiderivative of the function $\sqrt{x+7}$. Let $f'(x) = (x+7)^{1/2}$, then one of the antiderivatives is $f_1(x) = \frac{2}{3}(x+7)^{3/2}$, another is $f_2(x) = \frac{2}{3}(x+7)^{3/2} + 5$. Check $\frac{d}{dx} f_2(x) = \frac{d}{dx}\frac{2}{3}(x+7)^{3/2} + \frac{d}{dx}5 = \frac{3}{2}\frac{2}{3}(x+7)^{3/2-1} + 0 = \sqrt{x+7}$.

2. Since $2x$ is the derivative of x^2, then

$$\frac{2x}{\sqrt{1-x^2}} = \frac{-\frac{d}{dx}(1-x^2)}{\sqrt{1-x^2}}$$

$$= -2\frac{d}{dx}(1-x^2)^{1/2}.$$

So, an antiderivative of $\frac{2x}{\sqrt{1-x^2}}$ is

$$-2(1-x^2)^{1/2} + C,$$

where C is any constant.

3. What identities do you know involving $\cos^2(\theta)$? Think of $\cos(2\theta) = \cos^2(\theta) - \sin^2(\theta)$, then

$$\cos(2\theta) = 2\cos^2(\theta) - 1.$$

The problem is asking for a function of $y(x)$ such that

$$\frac{d}{dx}y(x) = \left(\frac{1}{2}\cos(2\sin(x)) + \frac{1}{2}\right)\frac{d}{dx}\sin(x).$$

The solution is: $y(x) = \frac{1}{4}\sin(2\sin(x)) + \frac{1}{2}\sin(x) + C$.

4. For the functions

   (a) $\frac{1}{x^2}e^{1/x}$, we know that

   $$\frac{d}{dx}e^{1/x} = -\frac{1}{x^2}e^{1/x},$$

   so an antiderivative of $(1/x^2)e^{1/x}$, would be

   $$-e^{1/x} + C,$$

   for any constant, C.

   (b) $2xe^{x^2}$, we know that $2x = \frac{dx^2}{dx}$, so an antiderivative of $2xe^{x^2}$ is

   $$e^{x^2} + C,$$

   for any constant C.

5. To find an antiderivative of $\cos^2(x)$, use the identity, $\cos(2x) = \cos^2(x) - \sin^2(x) = 2\cos^2(x) - 1$. If $y'(x) = \cos^2(x)$, then

   $$\frac{dy}{dx} = \cos^2(x) = \frac{1}{2}(1 + \cos(2x)), \text{ so,}$$

   $$y(x) = \frac{1}{2}x + \frac{1}{4}\sin(2x) + C.$$

6. For $\sqrt{1 - x^2}$ to be real, x must be between -1 and $+1$, therefore x is the sine of some number. Let $x = \sin(t)$. Treating t as a function of x, and differentiating implicitly with respect to x, we get $1 = \cos(t)\frac{dt}{dx}$. Let $y(x) = y(\sin(t)) = u(t)$. We are given

   $$\frac{dy}{dx} = \sqrt{1 - x^2}, \text{ or by the chain rule on } u$$

   $$\frac{dy}{dx} = \frac{du}{dt}\frac{dt}{dx},$$

   $$\frac{du}{dt}\frac{dt}{dx} = \sqrt{1 - x^2},$$

   $$\frac{du}{dt}\frac{1}{\cos(t)} = \sqrt{1 - \sin^2(t)} = \cos(t),$$

   $$\frac{du}{dt} = \cos^2(t),$$

   $$u(t) = \frac{1}{2}t + \frac{1}{4}\sin(2t) + C, \text{ from Problem 5.}$$

Replacing t with $\sin^{-1}(x)$, we have

$$y(x) = \frac{1}{2}\sin^{-1}(x) + \frac{1}{4}\sin(2\sin^{-1}(x)) + C, \text{ or}$$

$$y(x) = \frac{1}{2}\sin^{-1}(x) + \frac{2}{4}\sin(\sin^{-1}(x))\cos(\sin^{-1}(x)) + C,$$

$$y(x) = \frac{1}{2}\sin^{-1}(x) + \frac{1}{2}x\sqrt{1 - x^2} + C.$$

7. Notice that x^2 depends upon the derivative of $x^3 - 1$ in a simple way; so, an antiderivative of $x^2\sqrt{x^3 - 1}$ is

$$\frac{2}{9}(x^3 - 1)^{3/2} + C,$$

for any constant, C.

8. Antiderivatives of

(a) $\frac{2x^3}{\sqrt{1-x^4}}$. Since $2x^3$ depends on the derivative of $1 - x^4$ in a simple way, namely $\frac{d}{dx}(\frac{-1}{2})(1 - x^4) = 2x^3$, we can write

$$\frac{2x^3}{\sqrt{1 - x^4}} = (\frac{-1}{2})\frac{d(1 - x^4)/dx}{\sqrt{1 - x^4}}$$

$$= (\frac{-1}{2})\frac{d}{dx}\sqrt{1 - x^4}.$$

So an antiderivative of the given function is

$$\frac{-1}{2}\sqrt{1 - x^4} + C.$$

(b) $\frac{2x}{\sqrt{1-x^4}}$. Let

$$y'(x) = \frac{2x}{\sqrt{1 - x^4}},$$

and let $x^2 = t$. Then $\frac{dt}{dx} = 2x$. If $y(x) = y(\sqrt{t}) = u(t)$, then

$$\frac{dy}{dx} = \frac{du}{dt}\frac{dt}{dx}, \text{ so}$$

$$\frac{2x}{\sqrt{1 - x^4}} = \frac{du}{dt}2x.$$

Now, replacing x with $\sqrt{t}$, we have

$$
\begin{aligned}
\frac{du}{dt} &= \frac{1}{\sqrt{1-t^2}} \\
u(t) &= \sin^{-1}(t) + C, \\
y(x) &= \sin^{-1}(x^2) + C.
\end{aligned}
$$

9. Anti derivatives of

(a) $xe^x + e^x$, write this as

$$
x\frac{de^x}{dx} + e^x\frac{dx}{dx}
$$

which is the derivative of $xe^x + C$.

(b) $x\sin(x)$, find the derivative of $x\cos(x)$.

$$
\begin{aligned}
\frac{d}{dx}(x\cos(x)) &= \cos(x) - x\sin(x), \text{ or} \\
x\sin(x) &= \cos(x) - \frac{d}{dx}(x\cos(x)), \\
x\sin(x) &= \frac{d}{dx}\sin(x) - \frac{d}{dx}(x\cos(x)).
\end{aligned}
$$

Therefore, an antiderivative of $x\sin(x)$ is

$$
\sin(x) - x\cos(x) + C.
$$

10. Multiply and divide $\csc(x)$ by $\csc(x) + \cot(x)$, thus

$$
\csc(x) = \frac{\csc(x)\csc(x) + \csc(x)\cot(x)}{\csc(x) + \cot(x)},
$$

but the numerator is the negative of the derivative of $\csc(x) + \cot(x)$; that is,

$$
\csc(x) = \frac{-\frac{d}{dx}(\csc(x) + \cot(x))}{\csc(x) + \cot(x)}.
$$

Write this as the derivative of a logarithm.

$$
\begin{aligned}
\csc(x) &= -\frac{d}{dx}\ln(\csc(x) + \cot(x)) \\
&= \frac{d}{dx}\ln\left(\frac{1}{\csc(x) + \cot(x)}\right).
\end{aligned}
$$

Therefore an antiderivative of $\csc(x)$ is

$$\ln\left(\frac{1}{\csc(x)+\cot(x)}\right)+C.$$

11. Write $\frac{1}{x^3+1}$ as follows.

$$\frac{1}{x^3+1}=\frac{1}{(x+1)(x^2-x+1)}.$$

Now find constants A, B and D, such that

$$\frac{1}{(x+1)(x^2-x+1)}=\frac{A}{(x+1)}+\frac{Bx+D}{(x^2-x+1)}.$$

We get $A=1/3, B=-1/3$ and $D=2/3$, therefore,

$$\frac{1}{x^3+1}=\frac{1}{3}\left(\frac{1}{(x+1)}\right)+\frac{1}{3}\left(\frac{-x+2}{(x^2-x+1)}\right). \qquad \text{(I)}$$

Now let us write x^2-x+1 as $(x-\frac{1}{2})^2+\frac{3}{4}$, and write $-x+2$ as $-(x-\frac{1}{2})+\frac{3}{2}$, so the last fraction in Equation (I) can be written

$$\begin{aligned}\frac{1}{3}\left(\frac{-x+2}{x^2-x+1}\right) &= \frac{1}{3}\left(\frac{-(x-\frac{1}{2})+\frac{3}{2}}{(x-\frac{1}{2})^2+\frac{3}{4}}\right)\\ &= \frac{1}{3}\left(\frac{-(x-\frac{1}{2})}{(x-\frac{1}{2})^2+\frac{3}{4}}\right)+\frac{1}{3}\left(\frac{\frac{3}{2}}{(x-\frac{1}{2})^2+\frac{3}{4}}\right). \text{ (II)}\end{aligned}$$

Now multiply numerator and denominator of the last fraction in Equation II by $\frac{4}{3}$ getting

$$\frac{1}{3}\left(\frac{\frac{3}{2}}{(x-\frac{1}{2})^2+\frac{3}{4}}\right)=\frac{1}{3}\left(\frac{2}{(\frac{2x-1}{\sqrt{3}})^2+1}\right).$$

This makes Equation (I) look like this

$$\frac{1}{x^3+1}=\frac{1}{3}\left(\frac{1}{x+1}\right)-\frac{1}{3}\left(\frac{x-\frac{1}{2}}{(x-\frac{1}{2})^2+\frac{3}{4}}\right)+\frac{1}{3}\left(\frac{2}{(\frac{2x-1}{\sqrt{3}})^2+1}\right). \qquad \text{(III)}$$

The first term in Equation (III) is $\frac{1}{3(x+1)}$. This is $\frac{1}{3}\frac{d}{dx}(x+1)/(x+1)$. Hence, it is in the form of $u'(x)/u(x)$; in other words the derivative of the logarithm $\ln(u(x))$.

The second term is also in the form of a function divided into its derivative.

The third term is in the form of

$$\frac{2}{u^2(x)+1}.$$

This reminds us of the trigonometric identities involving the inverse tangent   That is, if $y(t) = \tan^{-1}(u(t))$, then $\tan(y(t)) = u(t)$, and the derivative is

$$
\begin{aligned}
\sec^2(y(t))y'(t) &= u'(t) \\
y'(t) &= \frac{u'(t)}{\sec^2(y(t))}, \\
&= \frac{u'(t)}{\tan^2(y(t))+1}, \\
&= \frac{u'(t)}{u^2(t)+1}.
\end{aligned}
$$

Here the $u(t)$ is $(2t-1)/\sqrt{3}$, and the $u'(t)$ is $2/\sqrt{3}$, which can be adjusted to a 2, by repositioning the factor $(1/\sqrt{3})$. Thus, we have the following antiderivatives for the terms in Equation (III).

$$
\begin{aligned}
\frac{1}{3}\frac{1}{x+1} &= \frac{1}{3}\frac{d}{dx}\ln(x+1) \\
\frac{1}{3}\frac{x-\frac{1}{2}}{(x-\frac{1}{2})^2+\frac{3}{4}} &= \frac{1}{6}\frac{d}{dx}\ln\left((x-\frac{1}{2})^2+\frac{3}{4}\right), \\
\frac{1}{3}\frac{2}{(\frac{2x-1}{\sqrt{3}})^2+1} &= \frac{\sqrt{3}}{3}\frac{d}{dx}\tan^{-1}(\frac{2x-1}{\sqrt{3}}).
\end{aligned}
$$

Therefore, an antiderivative of $1/(x^3+1)$ is

$$\frac{1}{3}\ln(x+1) - \frac{1}{6}\ln\left((x-\frac{1}{2})^2+\frac{3}{4}\right) + \frac{\sqrt{3}}{3}\tan^{-1}(\frac{2x-1}{\sqrt{3}}) + C.$$

Here is a check of this answer.

$$\frac{d}{dx}\left(\frac{1}{3}\ln(x+1) - \frac{1}{6}\ln\left((x-\frac{1}{2})^2 + \frac{3}{4}\right) + \frac{\sqrt{3}}{3}\tan^{-1}(\frac{2x-1}{\sqrt{3}})\right) = \frac{1}{x^3+1}$$

12. If

$$y'(x) = 1 + x^2 + \frac{x^4}{2!} + \frac{x^6}{3!} + ... + \frac{x^{2n}}{n!} + ...,$$

then

$$y(x) = C + x + \frac{x^3}{3} + \frac{x^5}{5 \times 2!} + \frac{x^7}{7 \times 3!} + ... + \frac{x^{2n+1}}{(2n+1) \times n!} + ...$$

20. Area Under a Curve

Answers to Exercises 20

1. The graph of $y = 1 - \cos(x)$ is shown in Exercise 20, Figure 1.

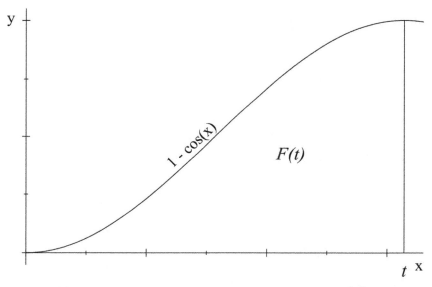

Exercise 20 Figure 1 Area under $y = 1 - \cos(x)$.

The area under the curve, $F(t)$ is an antiderivative of $1 - \cos(t)$; that is,

$$\begin{aligned} F'(t) &= 1 - \cos(t) \\ F(t) &= t - \sin(t) + C. \end{aligned}$$

385

The area under the curve between 0 and π is

$$F(\pi) - F(0) \;=\; \pi - \sin(\pi) + C - (0 - \sin(0) + C)$$
$$= \; \pi.$$

2. Graph of $1 + \sin(x)$. See Exercise 20, Figure 2

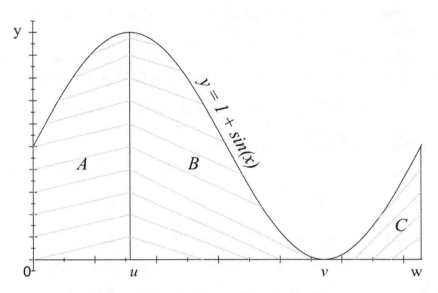

Exercise 20 Figure 2. $u = \pi/2,\ v = 3\pi/2,\ w = 2\pi.$

In each of the regions A, B, and C, the graph of $y = 1+\sin(x)$ is either increasing or decreasing throughout the region. So we may assume that the area is obtained by finding an antiderivative of $1 + \sin(x)$ in each region. Let $F(x) = x - \cos(x) + C$ be an antiderivative defining the area over the entire interval $[0, 2\pi]$. Then in the interval $[0, \pi/2]$, the area is

$$F(\tfrac{\pi}{2}) - F(0) \;=\; \frac{\pi}{2} - \cos(\tfrac{\pi}{2}) + C - (0 - \cos(0) + C),$$
$$= \; \frac{\pi}{2} + 1.$$

Similarly on the interval $[\pi/2, 3\pi/2]$

$$F(\tfrac{3\pi}{2}) - F(\tfrac{\pi}{2}) \;=\; \frac{3\pi}{2} - \cos(\tfrac{3\pi}{2}) + C - \left(\frac{\pi}{2} - \cos(\tfrac{\pi}{2}) + C\right),$$
$$= \; \pi.$$

and on the interval $[3\pi/2, 2\pi]$, the area is

$$F(2\pi) - F(\frac{3\pi}{2}) = 2\pi - \cos(2\pi) + C - \left(\frac{3\pi}{2} - \cos(\frac{3\pi}{2}) + C\right),$$
$$= \frac{\pi}{2} - 1.$$

Thus, the area from 0 to 2π is the sum of the three areas: $(\frac{\pi}{2} + 1) + \pi + (\frac{\pi}{2} - 1) = 2\pi$.

Alternatively, we could have just found the area by finding $F(2\pi) - F(0)$ all at once without finding the areas of the three regions.

$$F(2\pi) - F(0) = 2\pi - \cos(2\pi) + C - (0 - \cos(0) + C) = 2\pi.$$

3. The given function is $h(x) = 1 + x^2 + x^4/2$.

   The area under the curve between 0 and x is the antiderivative

   $$F(x) = x + x^3/3 + x^5/10 + C.$$

   Since $F(0) = 0$, then $C = 0$. So, the area between 0 and 3/2 is

   $$F(3/2) - F(0) = 3/2 + (3/2)^3/3 + (3/2)^5/10 - 0$$
   $$= 1083/320 \approx 3.384.$$

4. If $F'(x) = xe^{2x}$, find an antiderivative $F(x)$. Start by observing that e^{2x} is the derivative of $\frac{1}{2}e^{2x}$, so we write

   $$F'(x) = x\frac{d}{dx}(\frac{1}{2}e^{2x}).$$

   Let $u = x$, so $\frac{dx}{dx} = 1$, and $v = \frac{1}{2}e^{2x}$. Then the derivative of the product uv is

   $$\frac{d}{dx}(uv) = v\frac{du}{dx} + u\frac{dv}{dx}$$
   $$= \frac{1}{2}e^{2x} + xe^{2x}.$$

   Solve for xe^{2x}

   $$xe^{2x} = \frac{d}{dx}(uv) - \frac{1}{2}e^{2x}$$

or

$$xe^{2x} = \frac{d}{dx}(uv) - \frac{1}{4}\frac{d}{dx}e^{2x},$$
$$= \frac{d}{dx}\left(uv - \frac{1}{4}e^{2x}\right).$$

So, an antiderivative of xe^{2x} is

$$uv - \frac{1}{4}e^{2x} + C, \text{ or}$$
$$\frac{x}{2}e^{2x} - \frac{1}{4}e^{2x} + C.$$

5. Here $F'(x) = x\sin(x) = x\frac{d}{dx}(-\cos(x))$. Let $u = x$ and $v = -\cos(x)$, then differentiating the product uv,

$$\frac{d}{dx}(uv) = x\frac{d}{dx}(-\cos(x)) - \cos(x)$$
$$= x\sin(x) - \cos(x).$$

Solve for $x\sin(x)$

$$x\sin(x) = \frac{d}{dx}(uv) + \cos(x)$$
$$= \frac{d}{dx}(uv + \sin(x)).$$

An antiderivative of $x\sin(x)$ is

$$uv + \sin(x) + C, \text{ or}$$
$$-x\cos(x) + \sin(x) + C.$$

6. The graph of y is as shown in Exercise 20, Figure 3.

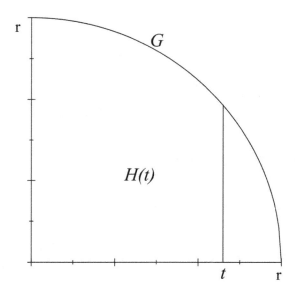

Exercise 20 Figure 3 Quarter Circle.

(a) Write the given $y = \sqrt{r^2 - x^2}$ as $y = r\sqrt{1 - \left(\frac{x}{r}\right)^2}$. For this square root to be real $\frac{x}{r}$ must be between -1 and $+1$. This means that $\frac{x}{r}$ is the sine of some number. Let $\frac{x}{r} = \sin(t)$, or $x = r\sin(t)$, and $y = r\cos(t)$.

$$\frac{dH(x)}{dx} = y = r\cos(t). \tag{I}$$

Using $H(x)$, we now define a function w of t.

$$H(x) = H(r\sin(t) = w(t).$$

Differentiate with respect to x, getting

$$\frac{dH}{dx} = \frac{dw}{dx}, \text{ so} \tag{II}$$

$$\frac{dH}{dx} = \frac{dw}{dt}\frac{dt}{dx}. \tag{III}$$

From the fact that $x = r\sin(t)$, we get

$$\frac{dx}{dx} = r\cos(t)\frac{dt}{dx}$$

$$\frac{dt}{dx} = \frac{1}{r\cos(t)}. \tag{IV}$$

Equations (III) and (IV) imply

$$\frac{dH}{dx} = \frac{dw}{dt}\frac{1}{r\cos(t)}. \tag{V}$$

From (I) and (V)

$$\frac{dw}{dt} = r^2\cos^2(t),$$
$$= r^2\left(\frac{1}{2} + \frac{1}{2}\cos(2t)\right).$$

So

$$w(t) = r^2\left(\frac{t}{2} + \frac{1}{4}\sin(2t)\right) + C,$$
$$w(t) = r^2\left(\frac{t}{2} + \frac{1}{2}\sin(t)\cos(t)\right) + C.$$

Now, from $w(t) = H(x)$, $r\sin(t) = x$, $t = \sin^{-1}(x/r)$, and $r\cos(t) = \sqrt{r^2 - x^2}$, we get

$$H(x) = r^2\left(\frac{\sin^{-1}(x/r)}{2} + \frac{1}{2}\frac{x\sqrt{r^2 - x^2}}{r^2}\right).$$

The area of the quarter circle is

$$H(r) - H(0) = r^2\left(\frac{\sin^{-1}(1)}{2} + 0\right) - r^2\left(0 + 0\right),$$
$$= r^2\frac{\pi}{4}.$$

(b) Therefore the area of the whole circle is πr^2.

7. The equation of the given ellipse is

$$\frac{x^2}{a^2} + \frac{y^2}{b^2} = 1.$$

Solving for y, the area we want to find is under the curve $y = b\sqrt{1 - x^2/a^2}$ from $x = 0$ to $x = b$, one quarter of the ellipse.

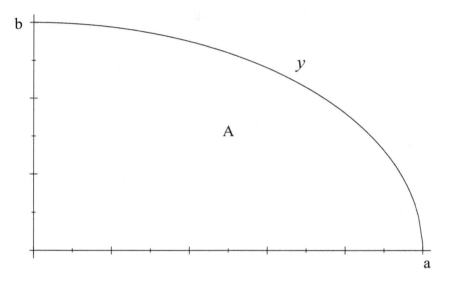

Exercise 20 Figure 4 Area under $y = b\sqrt{1 - x^2/a^2}$.

As before, $A(x)$ is the area under the graph from 0 to x, when

$$\frac{dA(x)}{dx} = y(x).$$

Note that since $y(x) = b\sqrt{1 - x^2/a^2}$, it can be real only if $\frac{x}{a}$ is between -1 and $+1$. Let $x = a\sin(t)$, then $y = b\cos(t)$. Define a function $w(t)$ as

$$A(x) = A(a\sin(t)) = w(t).$$

Differentiate with respect to x.

$$\frac{d}{dx}A(x) = \frac{dw}{dt}\frac{dt}{dx}.$$

But, from $x = a\sin(t)$, we get $\frac{dt}{dx} = \frac{1}{a\cos(t)}$, and from $\frac{dA}{dx} = y$, we can write

$$y = \frac{dw}{dt}\frac{1}{a\cos(t)}.$$

But $y = b\cos(t)$, so

$$\frac{dw}{dt} = ab\cos^2(t).$$

From Problem 6, we conclude

$$w(t) = ab\left(\frac{t}{2} + \frac{1}{2}\sin(t)\cos(t)\right) + C.$$

Replacing t by $\sin^{-1}(\frac{x}{a})$, and $w(t)$ by $A(x)$, we have

$$A(x) = ab\left(\frac{\sin^{-1}(x/a)}{2} + \frac{1}{2}\frac{x\sqrt{1 - x^2/a^2}}{ab}\right) + C.$$

Evaluating at the end points $x = 0$ and $x = a$, we get

$$\begin{aligned} A(a) - A(0) &= ab\left(\frac{\pi}{4} + 0 - (0 + 0)\right) \\ &= \frac{\pi ab}{4}. \end{aligned}$$

Therefore the area of the ellipse is πab.

8. The graph of $y = g(x)$ is broken into three regions all above the x-axis. If $H(x)$ is an antiderivative of $g(x)$, then in the first region A, where x goes from a to s, the area is $H(s) - H(a)$. The area in region B, over the interval $[s, t]$, is $H(t) - H(s)$, and in the last region, the area is $H(b) - H(t)$. Since area is additive for non-overlapping regions, then the area over the entire region from a to b is

$$H(s) - H(a) + H(t) - H(s) + H(b) - H(t) = H(b) - H(a).$$

9. When the graph is below the x-axis, we use the value of $-y$ to get a positive area. In the graph of $y = x^2 - 1$, graph is below the x-axis when x is between -1 and $+1$. Otherwise the graph is above the x-axis, that is on the intervals $[-2, -1]$ and $[1, 2]$. An antiderivative of y is

$$A(x) = \frac{x^3}{3} - x + C.$$

The area between the graph and the x-axis is

$$A(-1) - A(-2) - [A(1) - A(-1)] + A(2) - A(1) = 4.$$

21. Volume

Answers to Exercises 21

---

1. The equation of the graph G is $y(x) = \sqrt{r^2 - x^2}$ and $H(t)$ is the region under G when $0 \le t \le r$. When the region H is rotated about the x-axis then the circular cross section of the solid $S(t)$ at $x = t$ will have radius $y(t) = \sqrt{r^2 - t^2}$.

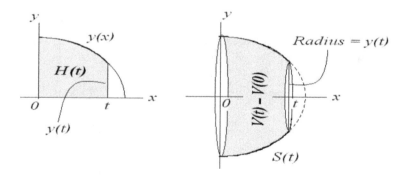

Exercise 21, Figure 1.

The area of that cross section will be $\pi \left(\sqrt{r^2 - t^2} \right)^2$. See Exercise 21, Figure 1. Therefore, the derivative of the volume $V(t)$ for any t in the

393

interval $[0, r]$ is

$$\frac{dV(t)}{dt} = \pi(r^2 - t^2).$$

So,

$$V(t) = \pi(r^2 t - \frac{t^3}{3}) + C.$$

Thus, the volume of $S(r)$ is

$$V(r) - V(0) = \frac{2}{3}\pi r^3.$$

This is the volume of the hemisphere. The volume of the sphere is $\frac{4}{3}\pi r^3$.

2. The ellipse equation is

$$\frac{x^2}{a^2} + \frac{y^2}{b^2} = 1.$$

Solving for y, as a function of x, we have

$$y(x) = \frac{b}{a}\sqrt{a^2 - x^2}.$$

When we rotate the ellipse about the x-axis, we get a solid with circular cross sections, the radii of which are $y(x)$ at each x in the interval $[0, a]$.

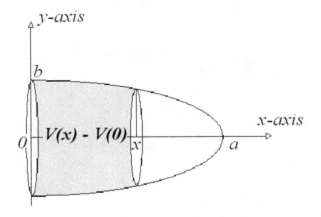

Exercise 21, Figure 2 Ellipsoid rotated about the x-axis.

The area of these circular sections are $\pi y^2(x)$, and the volume $V(x)$ from 0 to x satisfies the equation

$$\frac{dV(x)}{dx} = \pi y^2$$

$$= \pi \frac{b^2}{a^2}(a^2 - x^2).$$

An antiderivative is

$$V(x) = \pi \frac{b^2}{a^2}(a^2 x - \frac{x^3}{3}) + C.$$

The volume of one-half the ellipsoid is $V(a) - V(0)$

$$V(a) - V(0) = \pi \frac{b^2}{a^2} \frac{2}{3} a^3$$

$$= \frac{2}{3}\pi ab^2.$$

Therefore, the volume of the whole ellipsoid is $\frac{4}{3}\pi ab^2$.

3. To find the volume of the ellipsoid rotated about the y-axis, solve the equation in Problem 2, for x, and treat it as a function of y.

$$x(y) = \frac{a}{b}\sqrt{b^2 - y^2}.$$

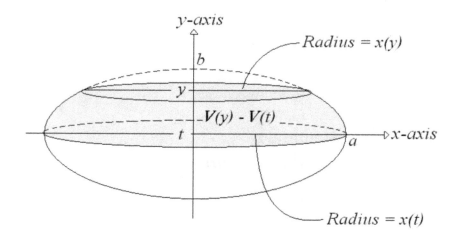

Exercise 21, Figure 3 Ellipsoid rotated about the y-axis.

The radii of the cross sections are $x(y)$, that is, functions of y. The areas of the cross sections are $\pi x^2(y)$.

$$\pi x^2(y) = \pi \frac{a^2}{b^2}(b^2 - y^2).$$

Let $V(y)$ be the volume for y in the interval $[0, b]$, then

$$\pi \frac{a^2}{b^2}(b^2 - y^2)(y - t) \leq V(y) - V(t) \leq \pi \frac{a^2}{b^2}(b^2 - t^2)(y - t).$$

Divide by $y - t$ and find the limit as $t \to y$, getting

$$\frac{dV(y)}{dy} = \pi \frac{a^2}{b^2}(b^2 - y^2),$$

and an antiderivative is

$$V(y) = \pi \frac{a^2}{b^2}(b^2 y - \frac{y^3}{3}) + C.$$

Therefore $1/2$ of the volume of this ellipsoid is $V(b) - V(0)$,

$$\begin{aligned} V(b) - V(0) &= \pi \frac{a^2}{b^2} \frac{2}{3} b^3 \\ &= \frac{2}{3} a^2 b \pi. \end{aligned}$$

The volume of the whole ellipsoid is $\frac{4}{3} a^2 b \pi$.

4. Assume that the tree trunk has a flat horizontal circle for its base. The wedge has triangular cross sections as shown in Exercise 21, Figure 4.

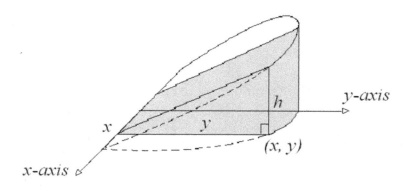

Exercise 21, Figure 4. The base y of each triangle is $\sqrt{r^2 - x^2}$.

Let the x-axis be the line through the center of the trunk's base. The origin is at the center of the trunk and the $30°$ vertices of all the triangles are on the x-axis. Then for x as shown in the figure, y will be $\sqrt{r^2 - x^2}$. This is the base of the triangle, and since it is a $30°/60°$ triangle the height is $\tan(30°)$ times the base. That is $h = b\tan(30°)$. Since the area of the triangle is $\frac{1}{2}bh$, then area $A(x)$ of the triangular cross section is

$$
\begin{aligned}
A(x) &= \frac{1}{2}b^2 \tan(30°) \\
&= \frac{1}{2}(r^2 - x^2) \tan(30°).
\end{aligned}
$$

For positive x and t with $x > t$, the volume $V(x) - V(t)$ satisfies the inequality

$$
\begin{aligned}
A(t)(x - t) &\leq V(x) - V(t) \leq A(x)(x - t) \\
A(t) &\leq \frac{V(x) - V(t)}{x - t} \leq A(x).
\end{aligned}
$$

The limit as $t \to x$, results in the following equation.

$$
\frac{dV(x)}{dx} = \frac{1}{2}(r^2 - x^2) \tan(30°).
$$

An antiderivative is

$$V(x) = \frac{1}{2}(r^2x - \frac{x^3}{3})\tan(30°) + C.$$

Then volume $V(r) - V(-r)$ is

$$V(r) - V(-r) = \frac{2}{3}r^3\tan(30°) = \frac{2\sqrt{3}}{9}r^3.$$

5. We sketch the graph depicting the solid obtained from two intersecting cylinders in Exercise 21, Figure 5. This drawing shows one-eighth of the solid; when we find its volume we will multiply by eight to get the volume of the whole solid.

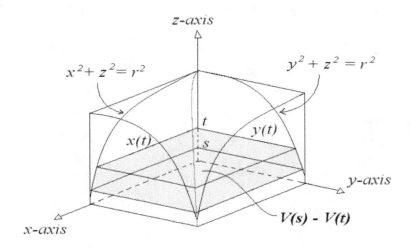

Exercise 21, Figure 5. Solid formed by intersecting right circular cylinders.

One cylinder has the x-axis as its axis and cuts the (y, z)-plane in a circle whose equation is $y^2 + z^2 = r^2$. The other cylinder has the y-axis for its axis and it cuts the (x, z)-plane in a circle whose equation is $x^2 + z^2 = r^2$. The solid cut by slices perpendicular to the z-axis at any value z forms a square whose sides are $y(z) = \sqrt{r^2 - z^2}$ and $x(z) = \sqrt{r^2 - z^2}$. Thus, if $z = t$ is any number on the z-axis between 0 and r, then $x(t)$ and $y(t)$ are both $\sqrt{r^2 - t^2}$. This makes the area of the square equal $r^2 - t^2$. If $0 \leq s \leq t \leq r$, and $V(t) - V(s)$ is the volume from $z = s$ to $z = t$, then the following inequality is true.

$$(r^2 - t^2)(t - s) \leq V(t) - V(s) \leq (r^2 - s^2)(t - s).$$

Dividing by $(t-s)$ and taking the limit as $s \to t$ gives us $\frac{dV(t)}{dt} = r^2 - t^2$. Therefore,

$$V(t) = r^2 t - \frac{t^3}{3} + C.$$

$$V(r) - V(0) = \frac{2}{3}r^3.$$

The volume of the intersections of the two cylinders is eight times $V(r)$;

$$8V(r) = \frac{16}{3}r^3.$$

22. Density and Pressure

Answers to Exercises 22

1. The point density $d(x, y, z)$ is $x^2 d$, where d is the density of the material in lbs/ft^3. The density does not vary in either the y or the z direction. For any y and z, the weight at a distance x from one end of the table is $F(x)$. For any numbers u and t, $0 \leq u < t \leq 5$, the weight of the slab between u and t is $F(t) - F(u)$. Since the maximum value of y is 3 and the maximum value of z is an unspecified number, say z_0, then $F(t) - F(u) \leq 3z_0(t - u)t^2 d$, which is the volume of the slab between u and t times the point density. Since $u < t$, then

$$3z_0(t - u)u^2 d \leq F(t) - F(u) \leq 3z_0(t - u)t^2 d.$$

As usual, we divide by $(t - u)$ and take the limit as $u \to t$, which gives us the derivative

$$\frac{dF(t)}{dt} = 3z_0 t^2 d,$$
$$F(t) = z_0 t^3 d + C.$$

Thus, from $t = 0$ to $t = 5$, the table top weighs $F(5) - F(0)$, or

$$F(5) - F(0) = 125 \times z_0 \times d \text{ pounds.}$$

2. Point density is $d(x, y, z) = xy^2 d$. The density does not vary in the z direction, and we are given that the maximum value of z is 4 inches, so $z_0 = \frac{1}{3} ft$. Let y be a fixed number and let u and t be two numbers on the x-axis with $0 \leq u < t \leq 5$. For a fixed y_0 the weight at any x

is $F_1(x)$, and the weight between u and t is $F_1(t) - F_1(u)$. Since $u > t$, this weight is less than $ty_0^2(t-u)z_0d$, and greater than $uy_0^2(t-u)z_0d$. That is,

$$uy_0^2(t-u)\frac{1}{3}d \;\le\; F_1(t) - F_1(u) \le ty_0^2(t-u)\frac{1}{3}d$$

$$uy_0^2\frac{1}{3}d \;\le\; \frac{F_1(t) - F_1(u)}{t-u} \le ty_0^2\frac{1}{3}d.$$

Taking the limit as $u \to t$, we have

$$\frac{dF_1(t)}{dt} \;=\; ty_0^2\frac{1}{3}d,$$

$$F_1(t) \;=\; t^2y_0^2\frac{1}{6}d + C.$$

Thus the weight $F_1(5) - F_1(0)$ at any fixed y_0 is

$$F_1(5) - F_1(0) = \frac{25}{6}y_0^2d \text{ pounds.}$$

Now let y vary from 0 to 3. That is y, is not just the fixed number y_0. The weight, depending on y is $F_2(y)$, and for any two numbers $u < t$ on the y-axis $F_2(t) - F_2(u)$ satisfies

$$\frac{25}{6}u^2(t-u)d \le F_2(t) - F_2(u) \le \frac{25}{6}t^2(t-u)d.$$

So

$$\frac{dF_2(t)}{dt} \;=\; \frac{25}{6}t^2d$$

$$F_2(t) \;=\; \frac{25}{18}t^3d + C.$$

The total weight is then

$$F_2(3) - F_2(0) = \frac{25}{18}27d \text{ pounds.}$$

3. Here the point density is $d(x, y, z) = x^2y^3z^4d \; lbs/ft^3$. The x values are in $[0, 5]$, y values in $[0, 3]$, z values in $[0, 1/2]$. For fixed y_0 and z_0, the

weight as a function of x is $F_1(x)$, and for two numbers $u < t$ on the x-axis,

$$u^2 y_0^3 z_0^4 (t - u)d \leq F_1(t) - F_1(u) \leq t^2 y_0^3 z_0^4 (t - u)d.$$

Dividing by $(t - u)$ and finding the limit as $u \to t$ yields

$$\frac{dF_1(t)}{dt} = t^2 y_0^3 z_0^4 d, \text{ so}$$

$$F_1(t) = \frac{t^3}{3} y_0^3 z_0^4 d + C.$$

Thus the weight for any fixed y_0 and z_0 is $F_1(5) - F_1(0)$

$$F_1(5) - F_1(0) = \frac{125}{3} y_0^3 z_0^4 d.$$

Now we let y vary. If $u < t$ are two numbers on the y-axis and $F_2(y)$ is the weight from 0 to y, then

$$\frac{125}{3} u^3 z_0^4 (t - u)d \leq F_2(t) - F_2(u) \leq \frac{125}{3} t^3 z_0^4 (t - u)d,$$

$$\frac{dF_2(t)}{dt} = \frac{125}{3} t^3 z_0^4 d$$

$$F_2(t) = \frac{125}{12} t^4 z_0^4 d + C.$$

Thus the weight for varying x and y density from $y = 0$ to $y = 3$, is

$$F_2(3) - F_2(0) = \frac{125}{12} 81 z_0^4 d.$$

Similarly if $F_3(z)$ is the weight using varying density on the x axis, we have

$$F_3(t) = \frac{125 \times 81}{60} z^5 d.$$

Now since z ranges from 0 to $1/2$, we have the weight of the table top is

$$F_3(1/2) - F_3(0) = \frac{125 \times 81}{60} (1/2)^5 d,$$

$$= \frac{675}{128} d \text{ pounds.}$$

4. Water weighs 62.2 lbs./ft.^{3}. The plane figure is the region whose boundaries are the downward pointing x-axis, the line $x = 1$, and the curve $y = \sqrt{x}$. At a depth of x, the force on the region is $F(x)$. If u and t are two numbers on the x-axis, $0 < u \leq t < 1$, then the force on the subregion between the depths u and t is $F(t) - F(u)$ and satisfies the inequality

$$62.2(t - u)u\sqrt{u} \leq F(t) - F(u) \leq 62.2(t - u)t\sqrt{t}$$

$$62.2u^{3/2} \leq \frac{F(t) - (u)}{t - u} \leq 62.2t^{3/2}.$$

The limit as $u \to t$ is

$$\frac{dF(t)}{dt} = 62.2t^{3/2}, \text{ so}$$

$$F(t) = \frac{2}{5}62.2t^{5/2} + C.$$

(a) The force on R is $F(1) - F(0) = 24.88$ pounds.

(b) The pressure is force per unit area. Here we compute the area at a depth of x as $A(x)$ by observing that between t and u, the area $A(t) - A(u)$ is between $\sqrt{u}(t - u)$ and $\sqrt{t}(t - u)$, so $\frac{dA(t)}{dt} = \sqrt{t}$, making $A(t) = \frac{2}{3}t^{3/2} + C$, so $A(1) - A(0) = \frac{2}{3}$ ft.^{2}. The pressure is

$$P = \frac{24.88}{2/3} = 37.32 \text{ pounds/square foot.}$$

5. The equation of the submerged circular gate as shown in Exercise 22 Figure 1 is $(x - 4)^2 + y^2 = 4$.

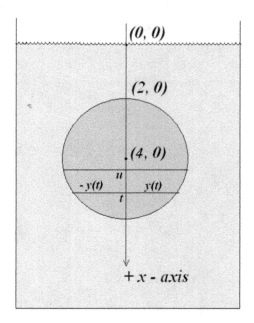

Exercise 22, Figure 1. Submerged vertical disk.

Solving for y as a function of x, we get

$$y(x) = \sqrt{4 - (x - 4)^2}.$$

We let the force on the disk at any depth x, $0 \leq x \leq 6$ be $F(x)$. If $t > u$ in the bottom half circle, then $y(u) > y(t)$ and $F(t) - F(u)$ will be between $t2y(u)(t-u)62.2$ pounds and $u2y(t)(t-u)62.2$ pounds because t is deeper than u. But the rectangle of width $(t - u)$ and length $2y(u)$ has a greater area than the $(t - u)$ by $2y(t)$ rectangle. That is

$$u2y(t)(t - u)62.2 \leq F(t) - F(u) \leq t2y(u)(t - u)62.2$$

or

$$u2y(t)62.2 \leq \frac{F(t) - F(u)}{t - u} \leq t2y(u)62.2.$$

Finding the limit as $u \to t$,

$$\frac{dF(t)}{dt} = 2 \times 62.2 t y(t).$$

Let $62.2 = w$, then since $y(t) = \sqrt{4 - (t-4)^2}$, we want an antiderivative of $2w \times t\sqrt{4 - (t-4)^2}$. Let $t - 4 = 2\sin(\theta)$, so

$$t \quad = \quad 4 + 2\sin(\theta), \tag{I}$$
$$y(t) \quad = \quad \sqrt{4 - (t-4)^2} = 2\cos(\theta) \tag{II}$$

From (I), we get

$$1 = 2\cos(\theta)\frac{d\theta}{dt}.$$

Write the function $F(t)$ as a function of θ as follows.

$$F(t) = F(4 + 2\sin(\theta)) = G(\theta) \tag{III}$$

Differentiating with respect to t,

$$\frac{dF(t)}{dt} = \frac{dG(\theta)}{d\theta}\frac{d\theta}{dt} \tag{IV}$$

Therefore,

$$\frac{dG(\theta)}{d\theta}\frac{1}{2\cos(\theta)} \quad = \quad 2w \times t \times y(t)$$
$$= \quad 2w \times (4 + 2\sin(\theta)) \times 2\cos(\theta)$$

This means that

$$\frac{dG(\theta)}{d\theta} \quad = \quad 16w \times \left(2\cos^2(\theta) + \sin(\theta)\cos^2(\theta)\right)$$
$$\frac{dG(\theta)}{d\theta} \quad = \quad 16w \left(\cos(2\theta) + 1 + \cos^2(\theta)\sin(\theta)\right).$$

We find an antiderivative to be

$$G(\theta) \quad = \quad 16w \left(\frac{1}{2}\sin(2\theta) + \theta - \frac{1}{3}\cos^3(\theta)\right) + C \tag{V}$$
$$= \quad 16w \left(\sin(\theta)\cos(\theta) + \theta - \frac{1}{3}\cos^3(\theta)\right) + C \tag{VI}$$

From (I), $\sin(\theta) = \frac{t-4}{2}$, so $\theta = \sin^{-1}(\frac{t-4}{2})$. From (II) we can recover $F(t)$.

$$F(t) = 16w\left(\frac{t-4}{2}\frac{\sqrt{4-(t-4)^2}}{2} + \sin^{-1}(\frac{t-4}{2}) - \left(\frac{4-(t-4)^2}{2}\right)^{3/2}\right) + C.$$

The same antiderivative holds for both the top half of the disk and the bottom half. For the top half we want $F(4) - F(2)$ and the bottom half we want $F(6) - F(4)$, which yields

$$F(4) - F(2) = 8\pi w - \frac{16w}{3} = (8\pi - \frac{16}{3})62.2 \text{ lbs., top half}$$

$$F(6) - F(4) = 8\pi w + \frac{16w}{3} = (8\pi + \frac{16}{3})62.2 \text{ lbs., bottom half.}$$

The area of each half disk is $2\pi\, ft^2$, so the pressure in the top half is $(4 - \frac{8}{3\pi})62.2\ lbs/ft^2$, and the pressure on the bottom half is $(4 + \frac{8}{3\pi})62.2\ lbs/ft^2$.

23. Arc Length

Answers to Exercises 23

1. Given $\sinh(t) = (e^t - e^{-t})/2$ and $\cosh(t) = (e^t + e^{-t})/2$, we can compute their derivatives from the derivatives of e^x.

   (a)

   $$\frac{d\sinh(t)}{dt} = \frac{d}{dt}\left(\frac{e^t - e^{-t}}{2}\right)$$
   $$= \frac{e^t + e^{-t}}{2} = \cosh(t).$$

   and

   $$\frac{d\cosh(t)}{dt} = \frac{d}{dt}\left(\frac{e^t + e^{-t}}{2}\right)$$
   $$= \frac{e^t - e^{-t}}{2} = \sinh(t).$$

   (b)

   $$\cosh^2(t) - 1 = \left(\frac{e^t + e^{-t}}{2}\right)^2 - 1$$
   $$= \frac{e^{2t} + 2 + e^{-2t}}{4} - \frac{4}{4},$$
   $$= \frac{e^{2t} - 2 + e^{-2t}}{4},$$
   $$= \sinh^2(t).$$

   Therefore, $\cosh^2(t) = 1 + \sinh^2(t)$.

(c) On the graph of $y = \cosh(t)$, the points $A = (-3, \cosh(-3))$ and $B = (3, \cosh(3))$ have arc length $\widehat{AB}$, and by the formula for arc length

$$
\frac{dS(t)}{dt} = \sqrt{1 + \left(\frac{d\cosh(t)}{dt}\right)^2}
$$
$$
= \sqrt{1 + \sinh^2(t)},
$$
$$
\frac{dS(t)}{dt} = \cosh(t).
$$

So

$$
S(t) = \sinh(t) + C.
$$

The arc length $\widehat{AB}$ is

$$
S(3) - S(-3) = \sinh(3) - \sinh(-3)
$$
$$
= \frac{e^3 - e^{-3}}{2} - \frac{e^{-3} - e^3}{2},
$$
$$
= e^3 - \frac{1}{e^3},
$$
$$
\approx 20.036.
$$

2. If G is the graph of $y = 2x^{2/3}$, find the length of an arc on G from $x = 1$ to $x = 8$. First, find $y'(x) = (4/3)x^{-1/3}$, then

$$
\frac{dS(x)}{dx} = \sqrt{1 + \left(\frac{4}{3}x^{-1/3}\right)^2},
$$
$$
= \frac{1}{3}x^{-1/3}\sqrt{9x^{2/3} + 16}.
$$

We know that $S(x)$ is going to be some multiple of

$$
(9x^{2/3} + 16)^{3/2}.
$$

Let

$$
S(x) = a(9x^{2/3} + 16)^{3/2}.
$$

To find a, we compute $dS(x)/dx$

$$
\frac{dS(x)}{dx} = \frac{3}{2}a(9x^{2/3} + 16)^{1/2}6x^{-1/3},
$$

implying that $9a = 1/3$ so, $a = 1/27$.

$$S(x) = \frac{1}{27}(9x^{2/3} + 16)^{3/2} + C.$$

The length from $x = 1$ to $x = 8$ is

$$S(8) - S(1) = \frac{1}{27}(104\sqrt{13} - 125).$$

3. If $y = \ln(\cos(x))$, then $y'(x) = -\tan(x)$, so

$$\begin{aligned}
\frac{dS(x)}{dx} &= \sqrt{1 + \tan^2(x)} \\
&= \sec(x).
\end{aligned}$$

Therefore,

$$S(x) = \ln(\sec(x) + \tan(x)) + C.$$

The length of the arc from $x = \pi/6$ to $x = \pi/4$ is

$$S(\frac{\pi}{4}) - S(\frac{\pi}{6}) = \ln\left(\frac{\sqrt{2}+1}{\sqrt{3}}\right).$$

4. For one loop of the sine curve, we use $y(x) = \sin(x)$, then $y'(x) = \cos(x)$ and

$$\begin{aligned}
\frac{dS(x)}{dx} &= \sqrt{1 + (y'(x))^2} \\
&= \sqrt{1 + \cos^2(x)}.
\end{aligned}$$

Write this as

$$\begin{aligned}
\frac{dS(x)}{dx} &= \sqrt{1 + 1 - \sin^2(x)} \\
&= \sqrt{2 - \sin^2(x)}, \\
&= \sqrt{2}\sqrt{1 - \frac{1}{2}\sin^2(x)}.
\end{aligned}$$

This is in the form of

$$\frac{dS(x)}{dx} = \sqrt{2}\sqrt{1 - k^2 \sin^2(x)}$$

with $k = \frac{1}{\sqrt{2}} \neq 1$. Hence, $S(x)$ would be in the form of an elliptic integral.

5. If $w(\theta) = \tan(2\theta)$, then $w'(\theta) = 2\sec^2(2\theta)$, and since $y'(\theta) = \sec^2(2\theta)$ then

$$2y'(\theta) = w'(\theta).$$

This means that we have, as antiderivatives, the following results

$$
\begin{aligned}
2y(\theta) &= w(\theta) + C, \\
y(\theta) &= \frac{1}{2}w(\theta) + \frac{1}{2}C, \\
y(\theta) &= \frac{1}{2}\tan(2\theta) + C_1.
\end{aligned}
$$

Here $C_1 = \frac{1}{2}C$.

24. Integral Notation

Answers to Exercises 24

1. Work done

   (a) The fifty pound bag leaks sand at the rate of $\frac{1}{7}$ lb. per foot raised. Therefore the force (here, it is weight) at any height x feet is $F(x) = 50 - \frac{1}{7}x$. So, the work done raising it 200 ft. is

   $$
   \begin{aligned}
   W(t) &= \int_0^t \left(50 - \frac{1}{7}x\right)dx \\
   W(t) &= \left.\left(50x - \frac{1}{14}x^2\right)\right|_{x=0}^{x=t}, \\
   W(t) &= 50t - \frac{1}{14}t^2.
   \end{aligned}
   $$

   So for $t = 200$, the work done is $\frac{100000}{14}$ ft.-lbs. Approximately 7143 ft-lbs.

   (b) If the bag had not leaked, the work done would have bee 50×200, or 10000 ft.-lbs.

2. Given $\int_a^u f(x)dx = \int_a^t f(x)dx + \int_t^u f(x)dx$, solving for $\int_a^t f((x)dx$, we get

   (a)
   $$
   \int_a^t f((x)dx = \int_a^u f(x)dx - \int_t^u f(x)dx.
   $$

413

(b) But if we consider defining the integral from a to u to t, we would get

$$\int_a^t f((x)dx = \int_a^u f(x)dx + \int_u^t f(x)dx,$$

so $-\int_t^u f(x)dx$ must be $+\int_u^t f(x)dx$.

(c) Since $\int_u^t f(x)dx = -\int_t^u f(x)dx$, then $\int_u^t f(x)dx + \int_t^u f(x)dx = 0$, or $\int_u^u f(x)dx = 0$, for any number u.

3. We use the property of sums

(a) For any $a_i \le b_i$

$$\sum_{i=1}^n a_i \le \sum_{i=1}^n b_i.$$

Therefore, if $f(x) \le g(x)$ on the interval $[a, b]$, then for any partition $\mathcal{P}_n[a, b]$,

$$\sum_{i=1}^n f(x_{i-1})\frac{b-a}{n} \le \sum_{i=1}^n g(x_{i-1})\frac{b-a}{n},$$

$$\lim_{n\to\infty} \sum_{i=1}^n f(x_{i-1})\frac{b-a}{n} \le \lim_{n\to\infty} \sum_{i=1}^n g(x_{i-1})\frac{b-a}{n},$$

$$\int_a^b f(x)dx \le \int_a^b g(x)dx.$$

(b) By the mean value theorem for integrals there is a number $z \in [a, b]$ such that

$$\int_a^b f(x)dx = f(z) \times (b - a).$$

Here, since $f(x) = c$ for all $x \in [a, b]$, $f(z) = c$. Hence,

$$\int_a^b cdx = c \times (b - a).$$

4. Let $a < b$. We consider the case where $f(x) \geq 0$ on $[a, b]$. Other cases can similarly be proved. Given $\int_a^b f(x)dx = A$; let $h = \frac{A}{b-a}$. Then $A = h \times (b - a)$. Therefore

$$\int_a^b f(x)dx = h \times (b - a).$$

If m is the minimum value of $f(x)$ on the interval $[a, b]$, then $m \leq f(x)$, for all $x \in [a, b]$, so from Problem 3(a),

$$\int_a^b m\,dx \;\leq\; \int_a^b f(x)dx, \text{ or}$$
$$m(b - a) \;\leq\; h(b - a).$$

Divide by $(b - a)$, getting $m \leq h$. Therefore h is not less than the minimum of $f(x)$.

5. We get the integrals from the sums having the following properties.

(a) From

$$\sum_{i=1}^{n} f(x_{i-1})\frac{b - a}{n} + \sum_{i=1}^{n} g(x_{i-1})\frac{b - a}{n} = \sum_{i=1}^{n} (f(x_{i-1}) + g(x_{i-1}))\frac{b - a}{n},$$

we get, after taking the limits as $n \to \infty$,

$$\int_a^b f(x)dx + \int_a^b g(x)dx = \int_a^b (f(x) + g(x))\,dx.$$

(b) If y is a constant, that is, y does not depend upon i, then

$$y\sum_{i=1}^{n} f(x_{i-1})\frac{b - a}{n} = \sum_{i=1}^{n} yf(x_{i-1})\frac{(b - a)}{n}.$$

Taking the limit as $n \to \infty$, gives us

$$y\int_a^b f(x)dx = \int_a^b yf(x)dx,$$

where y is any number that does not depend upon x.

6. Dummy variable equations for sums and integrals:

   (a) For any integer n,

$$\sum_{i=1}^{n} i^2 = 1^2 + 2^2 + +3^2 + \ldots + n^2, \text{ and}$$

$$\sum_{j=1}^{n} j^2 = 1^2 + 2^2 + +3^2 + \ldots + n^2.$$

   Notice that the $i's$ and the $j's$ are the same numbers so the sums are the same.

   (b) In general, if for each integer i and j, $a_i = a_j$, then

$$\sum_{i=1}^{n} a_i = \sum_{j=1}^{n} a_j.$$

   If the partitions $\mathcal{P}_n[a, b]$ and $\mathcal{R}_n[a, b]$ are as follows:

$$\mathcal{P}_n[a, b] = \{t_0, t_1, t_2, \ldots t_n\}$$
$$\mathcal{R}_n[a, b] = \{x_0, x_1, x_2, \ldots x_n\},$$

   and for each i the t_i's are equal to x_i's, then

$$\sum_{i=1}^{n} f(t_{i-1})\frac{b-a}{n} = \sum_{i=1}^{n} f(x_{i-1})\frac{b-a}{n},$$

   and the limits as $n \to \infty$ yield the following.

$$\int_a^b f(t)dt = \int_a^b f(x)dx.$$

7. To find the area of the shaded region, see Exercise 24 Figure 1. For the approximating sums, see Exercise 24 Figure 2.

   (a) The region A is the shaded region shown in Figure Exercise 24.1.

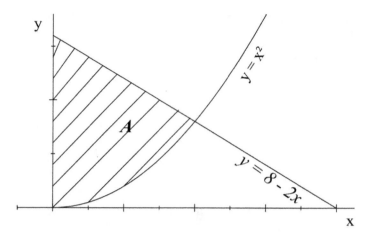

Exercise 24 Figure 1 The region A

(b) The partition and the approximating rectangles are shown in Exercise 24 Figure 2. The approximating sum is

$$\sum_{i=1}^{n}(8 - 2x_{i-1} - x_{i-1}^2)\frac{1}{2} = \frac{1}{2}\left(8 + \frac{27}{4} + 5 + \frac{11}{4}\right)$$
$$= \frac{45}{4} = 11.25.$$

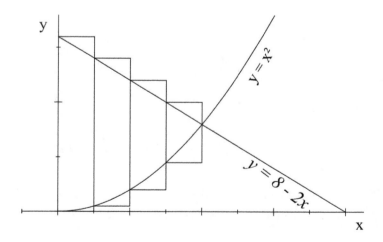

Exercise 24 Figure 2 The approximating rectangles.

(c) The exact area is obtained from

$$\int_0^2 (8 - 2x - x^2)dx \;=\; 8x - x^2 - \frac{x^3}{3}\Big|_{x=0}^{x=2}$$

$$=\; \frac{28}{3} = 9.333...$$

8. Differentiate, using the definition of derivative

(a) If

$$F(t) = \int_a^{t^2} f(x)dx,$$

then

$$F(r) \;=\; \int_a^{r^2} f(x)dx,\ \text{so}$$

$$F(t) - F(r) \;=\; \int_a^{t^2} f(x)dx - \int_a^{r^2} f(x)dx,$$

$$=\; \int_a^{t^2} f(x)dx + \int_{r^2}^{a} f(x)dx,$$

$$=\; \int_{r^2}^{t^2} f(x)dx.$$

But by the mean value theorem there exists a number z between t^2 and r^2, such that

$$\int_{r^2}^{t^2} f(x)dx = f(z)(t^2 - r^2).$$

Thus,

$$F(t) - F(r) \;=\; f(z)(t^2 - r^2)$$
$$\frac{F(t) - F(r)}{t - r} \;=\; f(z)(t + r).$$

Now as $r \to t$, $(t+r) \to 2t$, $z \to t^2$, and $\frac{F(t)-F(r)}{t-r} \to \frac{dF(t)}{dt}$, therefore

$$\frac{dF(t)}{dt} = 2tf(t^2).$$

(b) If a is a constant and $g(t)$ is a differentiable function on the interval $[a, b]$, we want to find $\frac{dF(t)}{dt}$, when

$$F(t) = \int_a^{g(t)} f(x)dx. \tag{I}$$

Let

$$F(r) = \int_a^{g(r)} f(x)dx,$$

then

$$F(t) - F(r) = \int_{g(r)}^{g(t)} f(x)dx.$$

By the mean value theorem for integrals, there exists z between $g(r)$ and $g(t)$, such that

$$
\begin{aligned}
F(t) - F(r) &= f(z) \times (g(t) - g(r)), \\
\frac{F(t) - F(r)}{t - r} &= f(z) \times \frac{g(t) - g(r)}{t - r}.
\end{aligned}
\tag{II}
$$

Taking the limit as $r \to t$ in Equation (II), we get

$$\frac{dF(t)}{dt} = f(g(t)) \times \frac{dg(t)}{dt}.$$

(c) Given

$$F(t) = \int_{\pi/2}^{t^2+t} \sin(x)dx,$$

we find that

$$\frac{dF(t)}{dt} = \sin(t^2 + t) \times (2t + 1).$$

Index